Diversified Cropping Pattern and Agricultural Development

Hasibur Rahaman

Diversified Cropping Pattern and Agricultural Development

A Case Study from Malda District, India

 Springer

Hasibur Rahaman
Department of Geography
Aligarh Muslim University
Aligarh, Uttar Pradesh, India

ISBN 978-3-030-55730-0   ISBN 978-3-030-55728-7   (eBook)
https://doi.org/10.1007/978-3-030-55728-7

This Springer imprint is published by the registered company Springer Nature Switzerland AG
The registered company address is: Gewerbestrasse 11, 6330 Cham, Switzerland

Preface

It is once said that agriculture is the backbone of an economy not because we derive food from it rather it provides livelihood options to many. This was true for most of the developed countries, but not now. On the contrary, this is still significant for most developing countries, for example India, as they place immense importance on overall development. The role of agriculture in India's socio-economic development has been changing from time to time. But the recent shifts in agricultural systems have experienced a new swing away from Green Revolution (1960) delusions. Existing literature suggests that Indian agriculture prior to 1960 was more eco-friendly, sustainable and diversified than what it is today. From 1960 to 2000, academic discussion on agricultural systems have been related with production and productivity of major crops. Recent issues that swing agricultural systems from being productivity based to sustainable agriculture are doubling farmer income, organic farming and most importantly diversification options. In my 5 years of research (2014–2019), I have assessed and explained the diversification opportunity as a future option for Indian agriculture through the crop diversification perspective. This book is based on my Ph.D. thesis completed in July 2019. Here, I wish to share research experience and contribute valuable insight to the body of knowledge.

The issues presented in this book are crop diversification and agricultural development which are very contemporary and relevant to promote sustainable agriculture especially in third world countries. Agricultural development in the climate change era, increasing farmer income, eco-friendly agriculture, employment generation, balanced food and food security, meeting the demand of growing consumerism in hotels and restaurants and others are such aspects that are directly linked with the present issue.

The aim of this book is to primarily help academicians, researchers and newcomers to study crop diversification and agricultural development. It includes methods to conduct research, techniques to measure development and diversification, tools to analyses development and diversification trends and interperate efficiency of development quantitatively, and policy linkage dimensions which are based on popular public response through qualitative analysis. Policy analysts, planners, project writers and decision makers may also find substantive materials derived from the Agriculture Census data.

This book is very Malda specific. The Malda pattern of development and diversification is somewhat unique in the sense that a vast majority of marginal and small land size rural farmers practice crop diversification measures with agricultural development. Yet, this pattern is somewhat similar to those in other districts, states and even countries as farmers do practice the same at small or large scale. But in case of others space, it is not similar process as marginal and small land size farmers to medium and large land groups. It is important to keep this difference in mind and make necessary policy and plans appraised by governments, private institutes and non-profit organizations while development pursuits for social and economic cause.

This book is organized into seven chapters, excluding the introduction and conclusion. The introductory part discusses research gap, research questions, objectives of the study, hypotheses, database and methodology of the study. Chapter 1 deals with physical, human and economic aspects of the study area with recent data. Chapter 2 provides a brief review of related literature, conceptual framework and theoretical underpinning as a prelude for the analysis of subsequent chapters. This is followed by Chapter 3, which examines the changing land use and cropping pattern by land size class wise of two decades, i.e., from 1995–96 to 2015–16. Chapter 4 analyses input and output determinants of agricultural development, their change and relation with crop diversification. Chapter 5 brings out the status of crop diversification across land size classes under different crop categories. Chapter 6 provides a detailed account of sample villages, development and diversification nexus, essential factors of such nexus and performance of agricultural development in selected villages. The final chapter assesses the future of agricultural development – problems and prospects and established policy linkage associated through qualitative assessment. The conclusion part ends with a summary of findings, suggestions and conclusions.

During this research endeavour, I received a lot of support, encouragement and guidance in the form of suggestions, comments and criticisms from my Ph.D. supervisor, faculty members, friends, family and well-wishers. I gratefully acknowledge the University Grants Commission (UGC), India, for providing funding during this research work. I am extremely grateful to Prof. Shamsul Haque Siddiqui for his valuable comments and suggestions.

Aligarh, U.P., India   Hasibur Rahaman
April 2020

Introduction

Nature of Problem

Presently, this world is in progress. The outcome of progress reflects in economic growth and development backed by scientific and technological innovations. Sectoral growth pushed by modern innovations has never been equal, rather keeps on changing. Changing contributions of the agricultural sector in overall economic growth have dwindled more apparently in the second half of the twentieth. According to the 2018 World Bank report, agriculture sits 3% of global gross domestic product (GDP) with 30% of global employees from 4% GDP to 31% employees in 2010. According to the Food and Agriculture Organization (FAO), still 60% of the global population who depend on agriculture and 80% of the young population who earn their living from it, as per analyses in 2016, are from rural areas. The performance of agriculture is even though not pleasing to global GDP but it remains important in poverty reduction, raising income, generating employment, ensuring food security, and supporting other sectors of the economy in this developing world.

The Indian economy has witnessed highly impressive rates of growth of about 6–9% per annum during the last 30 years. In fact, the economy has been growing fairly rapidly since the 1980s and breaking the pace of the Hindu rate (Basu 2010) of growth of about 3% recorded over a long period of 30 years since independence. Although overall growth of the economy has been impressive in ways, agricultural growth rate (1.5–2.1%) has been slowing down during the same periods, becoming a critical concern before policymakers, planners, and academia. Since, bulk of the poor live in rural areas, for higher growth rates of the economy to make any significant dent on their impoverishment, rural incomes have to grow rapidly too, a feat which can be achieved only through high rates of agricultural growth. The strategies and interventions needed to achieve high rates of agricultural growth and development have been the subject of interesting discussion in recent years. There have been policies, strategies and sponsored planned investment inducted to accelerate agricultural development at micro and macro level by both state and central government according to the nature of problems from time to time. Apart from diverse geographic environment, India's agriculture at present has been facing different problems such as climate change effect, natural calamities, reducing land size, population pressure, slow technological innovation and adaptations, use of low inputs, increasing price instability, volatile market, and, most importantly, slowing down public investment to overall development of agriculture. The nature and intensity of agricultural problems can be minimized for the sake of rural development if regional planning is given importance over macro level planning (Balasubramanyam 2019; Mohammad 1992).

In India, the idea of agricultural development became a topic of academic discussion only after the Green Revolution (1960s). Extended irrigation facilities, high yielding varieties of seeds (HYV), new implements and machinery, technological innovations, fertilizers, and pesticides transformed Indian agriculture from low subsistence to intensive subsistence cum slow commercial agriculture with sufficient and promising output. It was the need of the hour. At a cost of bumper production and productivity, soil sustainability drastically decreased, as a result, salinity and alkalinity became a new reality of the Green Revolution in India. The swelling drawback of the Green Revolution, increasing population, and new food security policy have been slowing down agricultural contribution in the national economy (Misra and Puri 2011).

The discourse of agricultural development includes two different aspects in its assessment process, namely inputs and outputs. Agricultural inputs include different infrastructure, technologies, and institutional aspects, whereas output parameters combine crop diversification, productivity, level of commercialization and specialization, and cropping intensity (Krishna 1992). Input dimensions of Indian agriculture have been changing from time to time as policy formulation targets any change in output dimensions. From 1960 to the post liberalization period (1992), increasing productivity and production were having main policy thrust. Therefore, crop specialization, mainly cereal-based cultivation, gained impetus during this period (Siddiqui 2006). During the same period, cropping intensity increased many folds, mainly of wheat and rice crop. On the other hand, commercial and plantation crops (jute and tea) recorded a small hike in cropping intensity and areal coverage.

The aim of the Green Revolution is achieved through productivity measures that continue until India becomes self-reliant in food crops. The swelling drawback of the Green Revolution along with increasing population, climate change, new food security policy, and challenges of sustainable agricultural orientation from international organizations like Food Agriculture Organization (FAO) have been slowing down agricultural contribution in the national economy (Misra 2011). Therefore, the post-liberalized agricultural policy adopts agricultural diversification and agricultural development by growing crops along with livestock which arise as new agricultural output dimension. After looking into the socioeconomic challenges of agricultural diversification, very soon policy implication turned it into crop diversification, as this requires lesser economic effort than agricultural diversification (Dorsey 1999). Agricultural development and crop diversification can go hand in hand as diversification promises many opportunities to farmers (Singh 2000) and best use of land in different seasons under high-value crops (De 2000). Profitability and risk of crop diversification require some policy concerns (Kumar et al. 2000) as diversification and agricultural development are essentially fostered by common institutional and infrastructural facilities (Acharya et al. 2011)

Crop diversification is not simply growing multiple crops at a time, rather it is considered a strategy to boost income, generate employment, minimize risk and uncertainties, improve food security, and preserve and augment natural resources, which ultimately leads to the development of the agricultural sector. Agricultural diversification especially of high-value crops is an encouraging option for marginal and small land size farmers with respect to its multifaceted opportunities (Vyas 1996; Joshi et al. 2004; Rao et al. 2006; De 2013; Siddiqui et al. 2016).

Crop diversification for developing countries offers many opportunities to boost the economic, social, environmental, and cultural life of people. A study on "Crop diversification, dietary diversity and agricultural income: Empirical evidence from eight developing countries" by Pellegrini and Tasciotti (2014) sheds light on the impact of crop diversification on nutrition and income of rural households from eight developing countries. The study has found that the majority of households grow food crops despite their low contribution to agricultural income, but certain households grow high-value crops (vegetables and fruits) which also show more dietary diversity than mere food crops–growing households do. In a study, Makate et al. (2016) found a strong correlation between crop diversification and crop income, and food security and crop productivity in Zimbabwe among smallholder farmers. The study also explains the perpetual positive impact of crop diversification on soil fertility, suppression of disease, improvement of agro-ecological systems, improvement of production stability, and crop yield income, which are perennially constant. In developing countries, smallholders mostly engage in food crop production for household consumption, which further amplifies their poor conditions. With small land size characteristics, if a diverse set of crops are grown, then probability of being poor and households remaining in poverty will decrease (Michler and Josephson 2017).

The aim of crop diversification to Indian farmers is mainly surrounded to economic benefits generated from high-value crops such as vegetables, floriculture and horticulture. A state-level study by Singh (2004) concluded that if horticultural crops are grown with proper strategy, then they have the potential to generate employment, income, export earnings, and good opportunities to achieve diverse nutritional security for small land size farmers. After looking at the many advantages of crop diversification, Hazra (2001) termed crop diversification as the new future of Indian farmers' income because it provides large-scale employment opportunities to rural India. Apart from employment, farmers practice crop diversification to improve soil quality, which ultimately increases overall farming profit (Johnston et al. 1995).

Crop diversification gets impetus from well-developed agricultural circumstances like irrigation infrastructure, technologies, insurance facilities, agricultural research and innovation, pro-farmer policy, intense cropping region, and others that favor the cultivation of high-value crops, which has resultantly enhanced income and livelihood options (Kumar and Gupta 2015). A block-level study on crop diversification by Siddiqui and Rahaman (2016) found that high agriculture devel-

opment leads to higher crop diversification. The dominant category of crop diversification is vegetables followed by flowers, spices, and fruits in the district. Among
many positives of crop diversification, if it is carried out properly in a given agricultural situation, then marginal and small land size farmers can be benefited through
income, foods, employment, and many others (Bhattacharyya 2008). A study on
"productivity enhancement of salt-affected environments through crop diversification" by Qadir et al. (2008) explained that with the help of advanced agricultural
technologies, potential crop diversification can increase in adverse ecological set up
like salt-affected regions (in saline water, coastal belt, saline soils, and salt
affected land).

Agricultural development and crop diversification have witnessed many opportunities which are mostly from crop-specific categories succeeded by market
demand. The diversification of horticultural crops (vegetables, fruits, and flowers)
increases significantly with particular agricultural advantages like arable land,
abundant labor, scientific production technologies, and access to market (Singh
2013). A study on "Diversification and its impact on smallholders: Evidence from a
study on vegetable production" by Joshi et al. (2006) concludes that vegetable cultivation in states like Uttar Pradesh is profitable on small farms as it requires household engagement. Therefore, income and employment ensue hand in hand. The
household's information further indicates that women in rural areas benefit more
from vegetable diversification in well-institutionalized agricultural setups. Beyond
vegetable crops, within food crops, nonfood crops and livestock farming diversification have increased nutritional security of small and marginal farmers (Bamji 2000).
The study further insists that the irrigation expansion, enabled credit opportunities,
application of new technologies, education, and optimum inputs management in
agriculture, especially in socially and economically backward regions, are driving
forces of crop diversification.

Agricultural development and induced diversification can potentially change
quality of life from different socioeconomic standards. A regional level study on
"Effect on Income and Employment of Diversification and Commercialization of
Agriculture in Kullu District of Himachal Pradesh" by Bala and Sharma (2005)
found that the introduction of vegetables (tomato, cabbage, pea, and cauliflower) as
a part of crop diversification has caused income hike and improved overall living
standards of farmers. The pace of change in life quality through high-value crop
diversification specially among marginal and sub marginal farmers is studied by
Chand (1996) in the Western Himalayan region. The study concludes that earlier
food grains met the needs of family consumption while crop diversification of high
value vegetable crops, pushed by irrigation, market, and motorable roads, increased
marginal land size output compared to bigger land size holders. And therefore, the
livelihood pattern of poor farmers has changed at a better pace than bigger farmers.
The livelihood pattern of farmers and farmworkers of crop diversification has experienced many social benefits which ultimately uphold personal life satisfactions
(Johnston 1995).

Social benefits of crop diversification are enormous. In developing countries,
malnutrition and child health are important policy concerns. To improve health and

child nutrition, crop diversification is an option to implement. A study on "Crop Diversification and child health: Empirical evidence from Tanzania" by Lovo and Veronesi (2019) has found positive effects on child height-for-age induced by increased dietary diversity. On the other hand, the dietary diversity at individual household level enhanced from production diversity increases nutritional intake of food in many developing countries (Sibhatu and Qaim 2018). A study on "Agricultural Diversity and Child Stunting in Nepal" by Shively and Sununtnasuk (2015) concludes that agricultural diversity (vegetables and root) is positively associated with the decrease of child stunting probability. The study also specifies the long-term nutritional benefits from agricultural diversification at household level. The nutritional benefits from crop diversification in the climate change era can also be used as a coping strategy to improve human health, dietary nutrition, and food security (Lovo and Veronesi 2019).

Food security is another social dimension of crop diversification. To get enough healthy and sufficient food for a hotter, drier, congested, and crowded future, the agricultural system must be resilient from all types of threat. Massawe et al. (2016) state "crop diversity is at the heart of global food security because it underpins the resilient agricultural system, while safeguarding the options for achieving food in future in a rapidly changing world". Schroth and Ruf (2014) state food security as an additional benefit of crop diversification. The diverse crops from young tree crops in humid tropical regions increase availability of nutrition at all times and also reduce the environmental vulnerability out of it. Pandey and Sharma (1996) conclude that agricultural diversification of Indian agriculture not only increases food grain production but also improves per capita food availability. For export-oriented agriculture, crop diversification further can be accelerated where the role of government may prove as catalyst. Dhawan et al. (1996) have examined the possibility of diversification of Indian agriculture from the viewpoint of food security. They showed that without a positive check on India's human and animal population, it would be difficult to diversify cereal crops oriented to others. In this context, Sharma et al. (1996) add that the area under cereal crops has decreased over time and the area is increasingly occupied by other high-value crops which foster food security at household level. A study on "Household food security, nutrition and crop diversification among smallholder farmers in the highlands of Guatemala" by Immink and Alarcón (1991) gives more holistic and refined views on food security and crop diversification through mixed cropping methods of humid regions. Crop diversification is a required measure for small landholders to ensure individual and household food security. The quality of diet can be guaranteed through consistent diversification to get short- and long-term nutritional food security.

Risk aversion due to crop failure or losses by natural calamities is an important advantage of crop diversification. Various studies spanning from state, district, and micro region pertaining to risk minimization exist (Agarwal 2004). The main findings stemming from these studies are listed as follows: first, the introduction of dairy and other allied activities along with regular crop cultivation have made farming profitable and efficient. Second, no doubt, the rural development programs provide an impetus to gain income and employment which fetch positively in diversification risk. Third, crop diversification also reduces risk of farm business as it allows crop rotation on a regular basis.

The factors or determinants of crop diversification and agricultural development are similar as agricultural growth and development push diversification through socioeconomic, infrastructural, institutional, technological, and policy intervention. Available literature also suggests that agricultural development induces diversification in different countries of the world. A wide range of empirical studies are available on the determinants of crop diversification and agriculture development. Chatterjee (2016) has found the importance of crop diversification, especially highly nutritious and high-value crops through water management. After analyzing the agricultural situation of water-scarce regions in different countries of Africa and India, states such as West Bengal (Bankura District) and Tamil Nadu (Annavasal village in the Kodavasal block in Thiruvarur District), the study suggests that water use efficiency can be increased significantly in Indian agriculture. And thereby, India's rural areas based on the local climate and soil type crop must be diversified in order to manage and save water as well. Dasgupta and Bhaumik (2014) have discussed "Crop Diversification and Agricultural Growth in West Bengal" for the period of 1980–1981 to 2009–2010 based on secondary data. Trends and patterns of crop diversification and their impact on agricultural development have been taken into consideration after inspecting the land holding situation of West Bengal. The study found that agricultural growth generally varies positively with crop diversification. The spread of technologies and infrastructural development still remain a major requirement. Besides, institutional measures are suggested by this study. Another study by Acharya et al. (2011) concludes that infrastructural and technological factors like fertilizers, market, proper irrigation, and road transportation are essential pre-requisites for crop diversification and agricultural development. Ghosh (2009) in his study found that access to socioeconomic and infrastructural facilities boosts better crop diversification and development. The study suggests higher investment for roads and transportation and innovation institutions. Shafi (2008) summed up that technological and institutional infrastructure like irrigation sources, machinery, HYV seeds, and use of fertilizers have made agriculture innovative and this innovation has a positive impact on crop diversification of different crops.

A study has been conducted by Viswanathan and Satyasai (1997) where they have examined different dimensions of commercialization and diversification in Indian agriculture. The study found that the states of Kerala and Andhra Pradesh recorded phenomenal shifts in the area from food grains to non-food grains while other parts like the western, northwestern, central, and the eastern region also displayed considerable shifts in terms of share in an area under non-food crops. This significant shift is mainly due to demographic factors, technology, infrastructure, policy environment, and global import demand. Kumar and Gupta (2015) in their study on "Crop Diversification towards High-value Crops in India: A State Level Empirical Analysis" examined trends in crop diversification towards high-value crops along with identification of major factors determining crop diversification from 1990–1991 to 2011–2012.

It is apparent from aforementioned studies that crop diversification may raise farm income and increase employment prospects in the rural sector. However, there are conflicting views regarding its effect on food security. The studies also indicate that infrastructure, technology, institutional measures, socioeconomic aspects, government policy, and research extensions are hypothesized to influence agriculture and crop diversification processes. While judging these findings, it may be noted that these studies are done in various micro-locations in the country or at the district or state levels using either primary household data or secondary data relating to different time periods. Extensive literature survey also confirms that there is no study on chosen aspects by land size classes category. This study fills this gap objectively. Since diversification and development are a continuing process and may change over time and across space, it is necessary to carry out such studies for different regions of the country using both aggregate data as well as household data. The present study is carried out in this broader framework on Malda district of West Bengal. It may be noted that there exists hardly any study on this subject for the district.

According to the 2015–2016 Agriculture Census, Malda district of West Bengal became the home of more than 96% of marginal and small farmers having land size less than two hectares. Intensive subsistence cum low commercial agriculture is only boosting their income and livelihood. The decreasing groundwater table, reduction in macro and micro soil nutrients, climate variability, erratic monsoon, low level of inputs and technological adaptation, instability and volatility, agriproduct price, and slowdown in public investments have been identified as major constraints in the district. In this context, the present work on "Diversified Cropping Pattern and Agriculture Development in Malda District" becomes an interesting issue of discussion.

Research Question

Being an important dimension of agricultural development, crop diversification is not simply growing multiple crops on the same field rather it offers farming sustainability with promising return. Therefore, in comparison to other dimensions of agriculture (commercialization and specialization), crop diversification assumes better agricultural output across different land size class categories over any region. Here is the research question:

In the present academic context, can agricultural development ensure crop diversification across different classified land size classes in Malda district?

Objectives

Present study is devoted to answering the research question raised by the author himself. To do so, six objectives and four hypotheses have been framed. The overall objective of the study is to examine the extent, pattern, factors, and future of crop diversification and its contribution to the development of agriculture in Malda district of West Bengal. More specifically, the objectives of the study are the following:

1. To examine the land use and cropping pattern change across land size classes
2. To analyze the impact of physical and non-physical determinants on agricultural outputs
3. To bring out the changing pattern of crop diversification across crop category under different land size classes at the block level
4. To measure the level and efficiency of agricultural development under land size classes at village level
5. To find out the major factors influencing agricultural development in different land size classes at the village level
6. To suggest evidence-based policy intervention for crop diversification and agricultural development in the district.

Hypotheses

The following four hypotheses set as null are to be tested for rejection under the assumption these are true:

1. **Ho:** There is no significant difference in the amount of change in cropping pattern across land size classes
2. **Ho:** Physical and non-physical determinants equally impact on agricultural outputs
3. **Ho:** Non-physical inputs negatively correlates with crop diversification
4. **Ho:** Agricultural inputs ensure efficiency of agriculture

Database

The present work is based on both primary and secondary sources of data.

Primary Data

Primary data was collected through different stages.

In the first stage, 15 villages have been carefully chosen for micro-level study. Here, one village is selected from each block. Only villages populated by maximum number of cultivators according to Census of India 2011 have been selected for the study. Purposive sampling technique is used to select 30 agricultural households from each village. These households include the farmer from marginal (< 1 hectare), small (1–2 hectare), semi-medium (2–4 hectare), medium (4–10 hectare), and large (> 10 hectare) land size category based on Agriculture Census of India. The representativeness (NSSO, Agriculture Census and Inputs Survey) is the only criteria to select 30 sample sizes. Thereby, a total of 450 households have been surveyed through schedule and questionnaire.

In the second stage (two phases), 15 Agricultural Development Officers (ADO) have been selected for interactive discussion and consultation on specific information about the block where they are posted. Information on strength, weakness, opportunities, and threats (SWOT) with respect to agriculture and development prospects has been obtained through questionnaires and discussion. In the second phase of the present stage, policy issues and development linkage related to agricultural development and crop diversification have been noted through participatory research appraisal (PRA).

In the third stage, an in-depth interview with Deputy Director of Agriculture (DDA Admin.) has been conducted to comprehend the selected theme in a more realistic way.

Secondary Data

Secondary data is obtained from the following sources:

1. Agriculture Census of India
2. Input Survey, Govt. of India
3. Census of India
4. National Sample Survey 59th Round, 2003
5. Statistical Handbook of Malda
6. Office of the DDA Admin., Malda
7. Office of the Horticultural Department, Malda
8. Office of the Soil Testing Laboratory, Malda
9. Comprehensive District Agricultural Plan (C-DAP)
10. Reports on Strategic Research and Extension Plan of Malda (SREP)
11. Agricultural Technology Management Agency (ATMA), Malda
12. District Irrigation Plan Book, Malda
13. Agricultural Report on Malda by Uttar Banga Krishi Vidyalaya (UBKV)

Apart from these, Human Development Report, Minority Concentrated District Reports, and others have been taken as concerned for required information.

In the study, two time periods data (1995–1996 and 2015–2016) on six-fold and size classes have been used. Up to March 2020, data of 2015–2016 and 2016–2017 were not published by the Agricultural Census and Input Survey; therefore, the said data have been extrapolated based on last Census data of 2010–2011 and 2011–2012. Before extrapolation, data have been interpolated from base year (last Census data) and then average of the last three years taken as actual data for 2015–2016 and 2016–2017. The technique of data interpolation is:

$$\textbf{Interpolation} = \frac{\text{Base year data} + (\text{Recent year data} - \text{Base year data})}{\text{Number years} (\text{here Five})}$$

Here, for Agriculture Census data, the base is 2005–2006 and the recent year is 2010–2011, while for input survey data, the base year is 2006–2007 and the recent year is 2011–2012.

For data split, the number of households and the gross cropped area have been taken as the base to the block.

Methodology

Qualitative and quantitative methods have been used for data analysis and data mining in the study.

Qualitative Techniques

1. Strengths, Weakness, Opportunities, and Threats (SWOT) analysis, a purely qualitative technique used in Chap. 7.
2. Participatory Rural Appraisal (PRA) is a qualitative-cum-quantitative technique applied for evidence-based policy suggestion in Chap. 7.
3. Qualitative Score Card ranking matrix is calculated for prioritized problem identification in Chap. 7.

Quantitative Techniques

1. The simple percentage technique has been used in Chaps. 1, 2, and 5.
2. To test first hypothesis of the study, the ANOVA test is used in the Second Chapter.
3. To compute standardized index for agricultural development, the UNDP method has been used in Chaps. 3 and 5.
4. To measure the impact of agricultural inputs on agricultural outputs, Ordinary Least Square (OLS) is used in Chap. 3.

5. To test significance level of framed hypothesis, the mean weightage index has been used in Chap. 3.
6. To calculate productivity, Yang's Yield Index is used in Chaps. 3 and 5.
7. Gibbs-Martin diversification technique is used in Chaps. 3 and 5.
8. To check degree of association, the Karl Pearson's correlation coefficient has been used in Chap. 3.
9. To determine breaking point of composite index and diversification index, the quartile technique is used in Chaps. 3 and 5.
10. Gibbs-Martin diversification, Simpson Diversification Index (SID), and crop counting method have been used to calculate crop diversification in Chap. 4.
11. The cropping intensity techniques are used in Chap. 5.
12. For factor reduction, the Principal Component Analysis (PCA) is used in Chap. 5.
13. To check efficiency of agricultural development in sampled villages, Data Envelope Analysis (DEA) has been used in Chap. 5.
14. To test a significant level of framed hypothesis two tailed T-test is used.

In the present study, data have been processed in Excel spreadsheet. And, aforementioned techniques and methods run in different software packages like Jamovi 1.20, R Studio, and Gretel software. QGIS 2.14 and Excel 16 have been used for maps and charts, respectively.

References

Acharya, S. P., Basavaraja, H., Kunnal, L. B., Mahajanashetti, S. B., & Bhat, A. R. (2011). Crop diversification in Karnataka: An economic analysis. *Agricultural Economics Research Review, 24*(2), 351–357.

Agarwal, I. (2004). Rationale of resource use through farm diversification. *Agricultural Economics Research Review, 17*(1), 85–100.

Alexandratos, N. (Ed.). (1995). *World agriculture: Towards 2010: An FAO study*. Rome: Food & Agriculture Org.

Arvis, J. F., Ojala, L., Wiederer, C., Shepherd, B., Raj, A., Dairabayeva, K., & Kiiski, T. (2018). *Connecting to compete 2018: Trade logistics in the global economy*. Washington, DC: World Bank.

Bala, B., & Sharma, S. D. (2005). Effect on income and employment of diversification and commercialization of agriculture in Kullu district of Himachal Pradesh. *Agricultural Economics Research Review, 18*(347-2016-16685), 261–269.

Balasubramanyam, V. N. (2019). *The economy of India*. London: Routledge.

Bamji, M. S. (2000). Diversification of agriculture for human nutrition. *Current Science, 78*(7), 771–773.

Bari, M. (2015). *Agriculture diversification and its impact on socio economic development in Aligarh district*. Doctoral dissertation, Aligarh Muslim University.

Basu, K. (2010). *The retreat of democracy and other itinerant essays on globalization, economics, and India*. London: Anthem Press.

Bhattacharyya, R. (2008, May). Crop diversification: A search for an alternative income of the farmers in the state of West Bengal in India. In *International conference on applied economics* (pp. 83–94).

Bruinsma, J. (Ed.). (2003). *World agriculture: Towards 2015/2030: An FAO perspective*. London: Earthscan.

Census, A. (2019). *All India report on number and area of operational holdings*. New Delhi: Agriculture Census Division Department of Agriculture, Co-Operation & Farmers Welfare Under Ministry of Agriculture & Farmers Welfare, Government of India. Retrieved from http://agcensus.nic.in/document/agcen1516/T1_ac_2015_16.pdf.

Census, A. (2020). *National Informatics Centre (NIC)*. Retrieved from Agriculture Census Division, DAC: http://agcensus.dacnet.nic.in/

Chand, R. (1996). Diversification through high value crops in Western Himalayan region: Evidence from Himachal Pradesh. *Indian Journal of Agricultural Economics, 51*(4), 652–663.

Chatterjee, S., (2016, January). Importance of crop diversification, *Economic and Political Weekly, 51*(16).

CSO (2007). *National Accounts Statistics 2007 and back issues*. Central Statistical Organization, Ministry of Statistics and Program Implementation, Government of India.

Dasgupta, S., & Bhaumik, S. K. (2014). Crop diversification and agricultural growth in West Bengal. *Indian Journal of Agricultural Economics, 69*.(902-2016-67970), 108–124.

De, U. K.. (2000). *Crop diversification in West Bengal from 1970–71 to 1994–95*. Unpublished thesis, Department of Economics, University of Burdwan.

De, U. K. (2013). Economics of crop diversification- an analysis of land allocation towards different crops. *Assam Economic Review, 8*(1), 9–29.

Dhawan, K. C., Singh, B., Prihar, R. S., Brar, S. S., & Arora, B. S. (1996). Diversification of Indian apiculture vis-a-vis food security. *Indian Journal of Agricultural Economic, 51*(4), 683–684.

Dorsey, B. (1999). Agricultural intensification, diversification, and commercial production among smallholder coffee growers in Central Kenya. *Economic Geography, 75*(2), 178–195.

Food and Agriculture Organisation. (2020). *Economic and social development department*. Retrieved from http://www.fao.org/economic/es-home/en/#.Xs4FRmgzZnJ: http://www.fao.org/home/en/

Ghosh, B. K. (2009). Factors affecting farmers decision to cultivate high-valued crops: A case study of Burdwan District of West Bengal. *IASSI Quarterly, 28*(1), 148–159.

GOI. (2006, December). *Towards faster and more inclusive growth: An approach to the 11th five year plan (2007–2012)*. New Delhi: Planning Commission, Government of India.

GOI. (2008–2009). *Agricultural statistics at a glance*. Ministry of Agriculture., http://agricoop.nic.in/

Hazra, C. R. (2001). In M. K. Papademetriou and F. J. Dent (Eds.). *Crop diversification in India. Crop diversification in the Asia-Pacific region* (pp. 32–50). Food and Agriculture Organization of the United Nations. Regional Office for Asia and the Pacific, Bangkok.

Immink, M. D., & Alarcón, J. A. (1992). *Household food security and crop diversification among smallholder farmers in Guatemala* (Food Nutrition and Agriculture, 4). Washington, DC: International Food Policy Research Institute.

Johnston, G., Vaupel, S., Kegel, F., & Cadet, M. (1995). Crop and farm diversification provide social benefits. *California Agriculture, 49*(1), 10–16.

Joshi, P. K., Gulati, A., Birthal, P. S., & Tiwari, L. (2004). Agriculture diversification in South Asia: Patterns, determinants and policy implications. *Economic and Political Weekly, 39*(24), 2457–2467.

Joshi, P. K., Joshi, L., & Birthal, P. S. (2006). Diversification and its impact on smallholders: Evidence from a study on vegetable production. *Agricultural Economics Research Review, 19*(347-2016-16776), 219–236.

Kothari, C. R., & Garg, G. (2014). *Research methodology methods and techniques*. New Delhi: New Age International (P) Ltd.

Krishna, G. (1992). The concept of agricultural development. In N. Mohammad (Ed.), *Dynamics development* (Vol. 7, pp. 29–36). New Delhi: Concept Publishing Company.

Kumar, S., & Gupta, S. (2015). Crop diversification towards high-value crops in India: A state level empirical analysis. *Agricultural Economics Research Review, 28*(2), 339–350.

Kumar, A. S. K., & Vashist, G. D. (2002, July–September). Profitability, risk and diversification in mountain agriculture: Some policy issues for slow-growing crops. *Indian Journal of Agricultural Economics, 57*(3), 356.

Lovo, S., & Veronesi, M. (2019). Crop diversification and child health: Empirical evidence from Tanzania. *Ecological Economics, 158,* 168–179.

Makate, C., Wang, R., Makate, M., & Mango, N. (2016). Crop diversification and livelihoods of smallholder farmers in Zimbabwe: Adaptive management for environmental change. *Springer Plus, 5*(1), 1135.

Massawe, F., Mayes, S., & Cheng, A. (2016). Crop diversity: An unexploited treasure trove for food security. *Trends in Plant Science, 21*(5), 365–368.

Michler, J. D., & Josephson, A. L. (2017). To specialize or diversify: Agricultural diversity and poverty dynamics in Ethiopia. *World Development, 89,* 214–226.

Misra, S. K., & Puri, V. K. (2011). *Indian economy* (p. 174). Mumbai: Himalaya Publishing House.

Mohammad, N. (1992a). *New dimensions in agricultural geography: Dynamics of agricultural development* (Vol. 7, p. 44). New Delhi: Concept Publishing Company.

Mohammad, N. (1992b). *New dimensions in agricultural geography: Spatial dimensions of agriculture* (Vol. 5, p. 71). New Delhi: Concept Publishing Company.

Panagariya, A. (2004). Growth and reforms during 1980s and 1990s. *Economic and Political Weekly, 39,* 2581–2594.

Pandey, V. K., & Sharma, K. C. (1996). Crop diversification and self-sufficiency in foodgrains. *Indian Journal of Agricultural Economics, 51*(4), 644–651.

Pellegrini, L., & Tasciotti, L. (2014). Crop diversification, dietary diversity and agricultural income: Empirical evidence from eight developing countries. *Canadian Journal of Development Studies/Revue canadienne d'études du développement, 35*(2), 211–227.

Qadir, M., Tubeileh, A., Akhtar, J., Larbi, A., Minhas, P. S., & Khan, M. A. (2008). Productivity enhancement of salt-affected environments through crop diversification. *Land Degradation & Development, 19*(4), 429–453.

Rao, P. P., Birthal, P. S., & Joshi, P. K. (2006). Diversification towards high value agriculture: Role of urbanization and infrastructure. *Economic and Political Weekly, 41*(26), 2747–2753.

Saha, P., & Basu, P. (2010). *Advanced Practical Geography-a Laboratory Manual* (Vol. 112, p. 435). Kolkata: Books & Allied (P) Ltd.

Sarkar, A. (2009). *Practical geography: A systematic approach.* New Delhi: Orient BlackSwan.

Sarkar, A. (2013). *Quantitative geography: Techniques and presentations* (p. 181). New Delhi: Orient BlackSwan.

Satyasai, K. J. S., & Viswanathan, K. U. (1996). Diversification in Indian agriculture and food security. *Indian Journal of Agricultural Economics, 51*(4), 674–679.

Schroth, G., & Ruf, F. (2014). Farmer strategies for tree crop diversification in the humid tropics. A review. *Agronomy for Sustainable Development, 34*(1), 139–154.

Shafi, S. P. (2008). *Diversified cropping pattern and agricultural development-a case study of Dadri Block, Gatuam Budh Nagar, UP.* Doctoral dissertation, Aligarh Muslim University.

Sharma, H. R. (2005). Agricultural development and crop diversification in Himachal Pradesh: Understanding the patterns, processes, determinants and lessons. *Indian Journal of Agricultural Economics, 60* (902-2016-68013).

Sharma, R. (2007). Agricultural development and crop diversification in Jammu and Kashmir: A district level study patterns, processes and determinants. *Review of Development and Change, 12*(2), 217–251.

Sharma, H. R. (2011). Crop diversification in Himachal Pradesh: Patterns, determinants and challenges. *Indian Journal of Agricultural Economics, 66* (902-2016-67882).

Sharma, R. K., Sharma, H. R., & Bala, B. (1996). Crop diversification and food security: An exploratory state wise analysis. *Indian Journal of Agricultural Economics, 51*(4), 680.

Shively, G., & Sununtnasuk, C. (2015). Agricultural diversity and child stunting in Nepal. *The Journal of Development Studies, 51*(8), 1078–1096.

Sibhatu, K. T., & Qaim, M. (2018). Meta-analysis of the association between production diversity, diets, and nutrition in smallholder farm households. *Food Policy, 77*, 1–18.

Siddiqui, S. H. (2006). *Environment, agriculture and poverty*. New Delhi: Concept Publishing Company.

Siddiqui, S. H. (2007). *Water resource and irrigation development in North Bihar plain* (Fifty Years of Indian Agriculture) (Vol. 2, pp. 258–265). New Delhi: Concept Publishing Company.

Siddiqui, S. H., & Rahaman, H. (2016, December). Crop diversification in relation to time and space: A study from Malda District. *International Journal of Informative & Futuristic Research*, (2), 04.

Singh, G. B. (2000, January). Green revolution in India–gains and pains. In *21st Indian geography congress*, Nagpur.

Singh, J. (2004). *The economy of Jammu & Kashmir*. Jammu: Radha Krishan Anand & Co.

Singh, R. P. (2013). Horticultural (high value agricultural) crops diversification in eastern India: II–employment opportunities and income generation strategies. *International Journal of Innovative Horticulture, 2*(1), 28–43.

Singh, J., & Sidhu, R. S. (2004). Factors in declining crop diversification: Case study of Punjab. *Economic and Political Weekly*, 5607–5610.

Survey, I. (2020). *National Informatics Centre (NIC)*. Retrieved from Agriculture Census Division, DAC: http://inputsurvey.dacnet.nic.in/

Swaminathan, M. (2010, January). Agricultural evolution during the last sixty years. *Yojana, 54*, 12–15.

The Government of India. (2012). *Indian agriculture at a glance- 2012* (pp. 259–260). New Delhi: Ministry of Agriculture.

The Government of India. (2013). *State of Indian agriculture 2012–13* (Vol. 2, p. 1). New Delhi: Ministry of Agriculture.

The Government of India. (2014). *Economic Survey of India 2013* (p. 174). New Delhi: Ministry of Finance.

Utpal Kumar De, U. K., & Chattopadhyay, M. (2010). Crop diversification by poor peasants and role of infrastructure: Evidence from West Bengal. *Journal of Development and Agricultural Economics, 2*(10), 340–350.

Viswanathan, K. U., & Satyasai, K. J. S. (1997). Fruits and vegetables: Production trends and role of linkages. *Indian Journal of Agricultural Economics, 52*(3), 574–583.

Vyas, V. S. (1996). Diversification in agriculture: Concept rationale and approaches. *India Journal of Agricultural Economics, 51*(4), 636–643.

World Bank. (2020). *Agriculture and food*. Retrieved from https://www.worldbank.org/en/topic/agriculture

Contents

List of Figures

List of Tables

Chapter 1
Overview of the Study Area

In the preceding introductory part, the nature of problems, research questions, objectives, hypotheses, database, and methodology have been discussed for meaningful shape of the study. The theoretical aspects of agricultural development and crop diversification are described on a systematic line to reach research questions and then fully fill objectives of the study. The theoretical aspects of the chosen subject have been analyzed with data in the present chapter entitled "Overview of the Study Area," i.e., Malda District of West Bengal. The analysis is presented in the following sequence. Next to the present section, a proper justification is given on why the author has selected this study area. Thereafter, a brief description of location, origin, and administrative divisions of the study area is presented and subsequently the aspects like physical, demographical, economical, market, and means of transportation have been interpreted. The maps and diagrams have been used to illustrate data in different sections.

1.1 Background of the District

1.1.1 Why Malda District?

In the present study, the Malda District of West Bengal have been selected as a study area because of two reasons; first, Malda is one of the most deprived districts in West Bengal in terms of Human Development Index (Malda District Human Development Report 2012) where about 87% of people live in rural area, out of which around 92% directly and indirectly depends on agriculture and allied activities (District Statistical Handbook 2015). Second, after the recommendation of Swaminathan Commission Report (2004–2006), the Government of India, under the aegis of Indian Council of Agricultural Research (ICAR), has selected 30 districts from different states of India to enhance crop diversification through agricultural development in modern techno-institutional setup. Malda district happens to be one of the districts in that list. Therefore, the present study is concerned to assess the impact of crop diversification program which ran in the district from 2005–2006 to 2012–2013. The impact evaluation of agriculture development is possible only after knowing all essential aspects of agriculture in the study area.

1.1.2   Location of the Study Area

The Malda District of West Bengal is a part of Jalpaiguri divisions, and it is located in the southernmost part of North Bengal. The district is bounded on southwest by river Padma, on north by the district of Uttar Dinajpur, and on southwest by Katihar district of Bihar. On the other side of the river Padma, Santhal Parganas district of Jharkhand state is situated. The district of Murshidabad in West Bengal is located in the south of Malda District. On the east, it is partly bounded by South Dinajpur district of West Bengal and Rajshahi district of Bangladesh.

The district is situated between $25°32'08''$ and $24°40'20''$ N latitudes and positioned entirely to the north of Tropic of Cancer ($23 \frac{1}{2}°$ N). The easternmost extremity of the district is marked by $80°28'10''$ E longitude and its westernmost extremity by $87°45'\ 50''$ E longitude. The area of district according to Surveyor General of India is 3719.223 km^2. The headquarter town of the district is English Bazar that is popularly known as Malda which is situated at $25°0'14''$ north latitude and $88°11'20''$ east longitude (Fig. 1.1).

1.1.3   Origin of the Name of the District

The name Malda was formerly applied to the town which is presently known as Old Malda. The present headquarter of the district is English Bazar which originated from the factory that was built by the East India Company to canyon trade mainly for silk and cotton textiles. The town which sprang up around the factory came to be known as Englezabad. Englezabad was converted to English Bazar in course of time at the beginning of the twentieth century.

The word Malda is derived from Arabic word, and therefore it is safe to conclude that the name Malda came into existence after the Muslim conquest of Bengal. The first mention of the name is found in the *Ain-i-Akbari* of Abul Fazl written during the reign of Emperor Akbar, but the name must have been existing before the Mughal conquest because there was no mention in *Ain-i-Akbari* that this particular name was conferred on the town of Old Malda after the Mughal conquest. The word "Mal" in Arabic means wealth, and the name Malda probably signifies that during the Muslim period, it was an important center of trade through which great wealth changed hands.

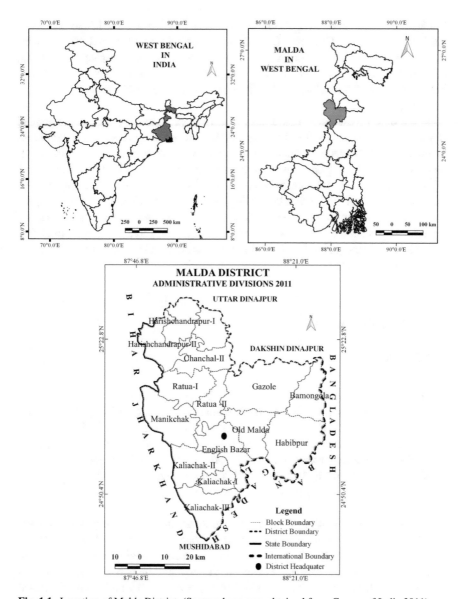

Fig. 1.1 Location of Malda District. (Source: base map obtained from Census of India 2011)

1.1.4 Administrative Divisions

The district has 15 blocks with English Bazar as an administrative center. For the sake of smooth administrative function, the district is divided into two sub-divisions, namely, Chanchal division and Sadar division. Chanchal division comprises four police stations, and these are rested with the responsibility to maintain law and order (under a political unit, i.e., block) of six blocks which are Harishchandrapur-I, Harishchandrapur-II, Chanchal-I, Chanchal-II, Ratua-I, and Ratua-II. The remaining nine blocks come under eight police stations in Sadar division which are Gazole, Bamangola, Habibpur, Old Malda, English Bazar, Manikchak, Kaliachak-I, Kaliachak-II, and Kaliachak-III. So, there is a total of 15 blocks ruled by 12 police stations (Table 1.1).

In the district, the apex institution for local governance is Malda Zila Parishad (District Council). Every development block corresponding to a local government institution is the respective panchayat samiti. At grassroots level there are 146 gram panchayats and 2008 gram sansads in the district, covering 1814 mouzas and 3733 villages that fall within the jurisdiction of Zila Parishad. There are 2 municipalities in the district, namely, Old Malda and English Bazar, with 18 and 25 wards, respectively. Urban governance function is done by municipalities. The district is primarily rural so that there is no statutory town. However, there are three small fast urbanizing clusters at Kendua, Kachu Pukur, and Ahio in Habibpur block (Primary Census Abstract 2011).

1.2   Physical Setting

1.2.1   Geology

Geologically, the district is recent in origin. Physically, the district is a part of the northern plain of India which is formed in quaternary geological period from alluvial-filled sedimentary rocks brought down by rivers to the south of Himalaya. The Malda gap which was created during tertiary geological period detached the peninsular plateau of India from eastern highlands which comprised Garo, Ghasi, and Jayantia in the state of Meghalaya. And this same Malda gap between Rajmahal hill and Garo hill gets filled with old and new alluvial soils during different geological ages.

The basement rock of the district suggests Archean granite complex on western part which seems to be the part of present Rajmahal hill. One drill hole at 210 m depth at Mandilpur of Manikchak block also supports of having granitic basement complex rock of Malda District. However, Archean granite is only found to a limited patch, and the base rock is formed from older alluvium which includes kankar and histolytica ferruginous concentration.

The surface soil of the district is made up of new alluvial deposits having two distinct nature and shows mild denudation of old alluvium. The northern part of district is formed from Himalayan piedmont (alluvial soil) which displays different physical and physiological character. The numerous streams from the Himalayas have brought down sediments and filled this part of district. The recent alluvium of surface soil of the district displays dark loose to reddish brown color with well-oxidized argillaceous bed having high water content. Apart from these, river Ganga

Table 1.1 Administrative units of Malda District

Sub-division	Police stations	C.D. Block/M	Area in sq. km	No. of households	Rural				Urban		
					Village	Gram sansad	Gram panchayat	Panchayat samiti	Municipality		Census town
									Number	Ward	
Chanchal	Harishchandrapur	Harishchandrapur-I	171.41	44,284	104	121	7	1	–	–	–
		Harishchandrapur-II	217.21	49,311	73	141	9	1	–	–	–
	Chanchal	Chanchal-I	162.14	49,273	98	135	8	1	–	–	1
		Chanchal-II	205.22	43,218	90	119	7	1	–-	–	–
	Pukhuria	Ratua-I	230.53	56,241	95	166	10	1			–
	Ratua	Ratua-II	173.93	43,168	48	124	8	1	–	–	–
Sadar	Gazole	Gazole	513.65	75,068	286	220	15	1	–	–	2
	Bamangola	Bamangola	205.91	32,154	141	102	6	1	–	–	–
	Habibpur	Habibpur	396.07	47,951	233	149	11	1	–	–	4
	Malda	Old Malda	215.66	33,629	112	97	6	1	–	–	3
		Old Malda (M)	9.00	16,479	–	127	9	1	1	18	1
	English Bazar	English Bazar	251.52	58,815	108	–	–	–	–	–	3
		English Bazar (M)	13.63	42,867	–	204	14	1	1	25	1
	Manikchak	Manikchak	321.77	59,567	72	169	11	1	–	–	–
	Kaliachak	Kaliachak-I	105.37	80,508	49	–	–	–	–	–	11
	Baishnabnagar	Kaliachak-II	222.73	44,913	40	162	11	1	–	–	1
		Kaliachak-III	260.12	69,545	65	232	14	1	–	–	2
District Total			3733	846,991	1613	2268	146	15	2	43	29

Source: District Statistical Handbook of Malda District 2015

and Mahananda also show an active role to form fresh alluvium during monsoon floods each year. The numerous small streams renewed surface soil each year as the district received a good amount of monsoon rain. The thickness of the soil is not even, rather it is a mixed composition of silt, gravel, and clay which are found almost similar throughout the district.

1.2.2   Physiography

There are three distinct physiographic divisions in the district. The Mahananda river flows from north to south all through the middle of the district. The eastern side of Mahananda river known as "Barind" is formed from older alluvial and red clay soil. The eastern part of Mahananda river is divided into two parts by Kalindri river which is the tributary of river Ganga flowing from west to east. The north of Kalindri river is known as "Tal" region which is formed of new alluvial soil composed of dark green and yellowish weathered soil. The south of Kalindri river is identified as "Diara" formed from new and fresh alluvium drained by river Ganga. A detailed description of these divisions is presented in Table 1.2.

There are no known mountains and hills in the district. A few elevated tracts and highlands have been found with ranging altitude from 15 to 35 meters above the level of river Ganga in Barind and Tal region. The highest elevation above mean sea level is found at Pandua in Gazole block. Some dissected tracks with altitude extending from 30 to 40 meters are also found in Bamangola, Habibpur, Chanchal-II, and Kaliachak blocks. The river course suggests that the gradual slope of the district is from north to south (Fig. 1.2).

1.2.3   River Systems

The principal rivers of the district enumerating from the east are the Punarbhava, the Tangon, the Mahananda, the Kalindri, and the Ganga. The river Ganga, which forms the southwestern boundary, receives water from all other rivers of the district. A detailed account of river systems of Malda District is displayed in Fig. 1.3.

Punarbhava
The Punarbhava river leaves the district of South Dinajpur and forms a few miles boundary between the police stations of Tapan (South Dinajpur district) and Bamangola (in Malda District). From these noted places, this throws outside channels at mouzas Mahadevpur in Khutadha police station of Bamangola block. These two side channels join together and combine a stream which is known as Haria river. It is a tributary of Mahananda which joins Bangladesh. In the district, the total length of Punarbhava including its tributaries is nearly 64.4 km. The occurrence of flood in this river is associated with heavy monsoon and overflow in Mahananda river.

Table 1.2 Block-wise physiographical characteristics

Region	Characteristics	Block covered	Area (in ha)
Tal	The average elevation of this region is low compared to the other two regions; therefore during the rainy season, this region is inundated very often. The general orientation of the slope of this region is from north and north-east to southwest. The nature of soil varies from clay to clay loam type with heavy texture. The textural classes of soil are sandy (2–3%), sandy loam (8–9%), loam (35–36%), clay loam (42–43%), silt loam (4–5%), and clay (5–6%). The main crops grown here are paddy, wheat, pulses, oilseeds, vegetables, makhana, and jute	Ratua-I Ratua-II Chanchal-I Chanchal-II Harishchandrapur-I Harishchandrapur-II	**114,099**
Diara	This is the flatland on the western side of Mahananda. It has light soils of new alluvium type. The water level is generally high and within the reach of shallow tube well even in summer. The soil textures of this region are sandy (7–8%), sandy loam (15–16%), loam (45–47%), clay loam (13–156%), silt loam (11–12%), and clay (5–6%). The major crops are paddy, vegetables, horticulture, jute, wheat, pulses and oilseeds	English Bazar Manikchak Kaliachak-I Kaliachak-II Kaliachak-III	**112,188**
Barind	This region is located comparatively on higher altitude (highest altitude 69.7 m in Gazole) than Tal and Diara. The area expands from east of Mahananda river and is characterized by extensive undulation and successive ridges with lowland topography having small water sources. The ground is hard. The groundwater table is not accessible easily except in the rainy season. The nature of soil is old heavy alluvium. Textural classes of soil are sandy loam (0–5%), loam (10–12%), clay loam (30–35%), and clay (55–57%). The main crops are Kharif rice, jute, and autmm rice	Gazole Bamangola Habibpur Old Malda	**132,851**

Source: CDAP XII plan of Malda District 2015

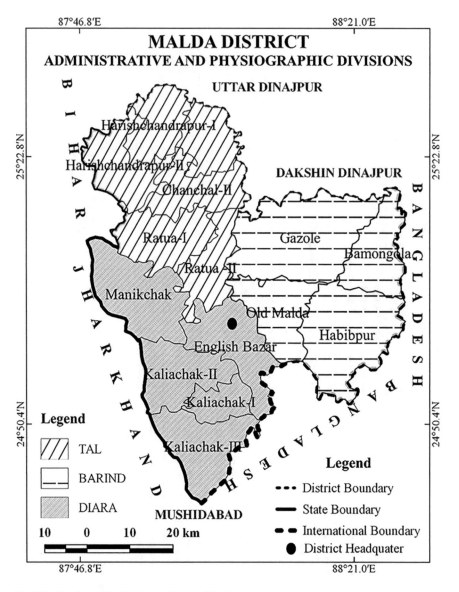

Fig. 1.2 Physiographic divisions of Malda District

Tangan

The river Tangan enters in the district from the north. The river makes a natural boundary between Gazole and Bamangola blocks, and further south, it forms a boundary between Habibpur and Malda police station before it joins Mahananda river. This river runs through Gazole, Habibpur, Bamangola, and Old Malda blocks.

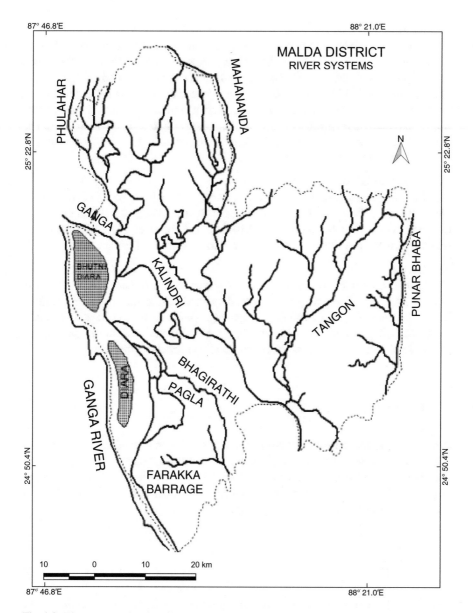

Fig. 1.3 River systems in Malda District

The length of the river Tangan is nearly 64.6 km in the district. There are two tributaries, namely, Mara Tangon and Chunakhali Kahl. Since, the river Tangan falls into Mahananda river (main river), the flood in it is associated with the main river.

Mahananda

The Mahananda river enters into the district from the north at the trijunction of Chanchal-I, Ratua-II, and Gazole blocks. It makes a boundary between Ratua and Gazole police stations and streams across English Bazar and Old Malda blocks. Slightly north of Old Malda block, this river is joined by river Tangan at Aiho. From north to south, this river passes over four blocks, namely, Gazole, Habibpur, Old Malda, and English Bazar, and touches the boundary of Ratua-II and Chanchal-I. The length of this river including two tributaries (Kalindri and Pagla) is 88.6 km. It ultimately joins the Ganga in Bangladesh.

Kalindri or Fulahar

The Kalindri river enters into the district from west at Mihaghat in Harishchandrapur-II block. Before falling into Mahananda river at Bachamari of Old Malda block, the river passes over Harishchandrapur-II, Manikchak, Ratua-I, and English Bazar. This river also known as Fulahar in Katihar and Purnia districts of Bihar state.

The Ganga

The Ganga flows from west and enters into the district at Gadai Char of Bhutni in Manikchak block. The river passes over Manikchak, Kaliachak-II, and Kaliachak-III blocks and joins three tributaries, namely, Fulahar, Bhagirathi, and Kalindri rivers. Including major tributaries, the length of the river is 172 km in the district. The end point of this river is at Pardeonapur in Kaliachak-III block.

1.2.4 Beels (Ponds)

Apart from the abovementioned rivers, the district has many small and big wetlands which are locally known as beels or ponds. During monsoon season, the low, depressed, and subsidized parts get filled with water, and these become the sources of pisciculture and wet farming in the district. The major concentration of beels is found in the Tal region due to its physiographic advantages.

1.2.5 Soil

The soil is the natural medium for growth of healthy plants under favorable conditions. The nature and type of soils determine which crop is to be planted that ensures better productivity. The clay soil with adequate water supply supports wetland crop cultivation like paddy, jute, makhana, water lily, and others. Sandy soil is rich in iron content, but only few crops like watermelon, cucumber, and few dry cactus trees grow. The loamy soil is best for agricultural purposes. Most crops with adequate irrigation and fertilizer are grown in loamy soil. The sedimentary-based alluvial soil is favorable for agricultural purposes. The study area is bestowed with Indo-Gangetic alluvial soil. And therefore, most cereals, pulses, oilseeds, vegetable, and horticulture crops get sufficient irrigation bases here.

1.2.5.1   Soil Types

There are several factors which are responsible for the development of different kinds of soils with variable texture, structure, color, depth, horizon, concretions, etc. According to the broad classification of soils, the Malda District has three kinds of soils, namely, (i) red soils (77,700 ha), (ii) Vindhya alluvium soils (17,130 ha), and a major part the (iii) Ganga alluvium soils (243,540 ha).

In the district, sandy soil covers around 3%, while coarse-loamy soils including sandy loam, loam, and silt loam textural classes spread over 12% in the area. Fine loamy soils encompassing sandy clay loam, clay loam, and silty clay loam textural classes occupy another 37.8% of the area. Fine soils including sandy clay, silty clay, and clay texture inhabit 37.1% of the area. Based on soil texture, there are as many as six types of soil found in the district (Fig. 1.4).

1.2.5.2   Soil Nutrition

Based on quality and chemical composition of soils, the soil nutrients are of two types, namely, macronutrients and micronutrients. For healthy growth and production of crops, the optimum amount of nutrients is required. Most of the nutrients naturally occur in soil. The unscientific agricultural practices have disturbed nutritional balance of soil; hence fertile soil in due course of time has become uncultivable. For economically profitable agriculture, well-maintained and balanced nutritious soil is essential.

Soil Health Card survey has identified as many as ten macro and micro soil nutrients to know the soil quality status from village to national levels. The essential macronutrients are nitrogen (N), phosphorus (P), and potassium (K) combined in a

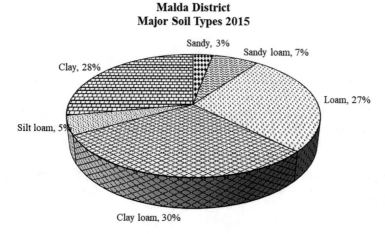

Fig. 1.4 Major soil types in Malda District. (Source: data from CDAP XII plan, Malda District 2015)

Table 1.3 Soil nutrient deficiencies in Malda District 2018–2019

Block	Macronutrients				Micronutrients					
	N	OC	P	K	Zn	Fe	Cu	Mn	B	S
Bamangola	69.8	65	8.69	9.48	96.5	5.43	12.9	33.7	98.6	100
Chanchal-I	80.2	78.9	0.68	12.4	89.1	2.43	13.6	33	99.5	99.8
Chanchal-II	73.3	73.2	0.39	8.58	94	8.46	8.62	29.2	99	99.2
English Bazar	72.9	71.6	1.51	8.05	91.9	3.37	8.1	37.2	99.8	100
Gazole	64.6	66.3	14.1	10.7	92.6	37.9	15.2	26.5	99.6	97
Habibpur	67.4	67.8	1.91	8.55	95.2	5.63	5.62	38	99.5	99.8
Harishchandrapur-I	81.3	81.9	0.26	11.8	95.8	0.36	7.41	32.8	99.2	100
Harishchandrapur-II	80.5	79.3	0.49	13	93.6	0.2	7.51	30.7	99.8	100
Kaliachak-I	63.3	61.5	0.9	7.76	92.6	5.45	16	34.1	98	99.5
Kaliachak-II	77.1	77.8	3.46	6.38	85.1	0.69	8.35	28.7	99.8	100
Kaliachak-III	74.2	73.5	1.09	8.85	92.4	6.03	12.9	40.9	99.5	100
Manikchak	76.4	77.5	4.52	10.2	89.3	14.6	9.92	31.6	99.3	98.7
Old Malda	69.7	69.1	1.24	8.26	95.4	10.4	4.05	29.6	99.9	99.7
Ratua-I	77.2	75.8	0.89	9.59	83.3	0.6	11.4	23.7	99.6	100
Ratua-II	77	77.5	1.36	9.45	93.5	12.1	11.1	40.7	98.3	99.5
District	73.3	72.9	3.59	9.76	92.1	9.7	10.1	32.5	99.3	99.4

Source: data from Soil Health Card Report 2018–2019

ratio of 4:2:1. The other micronutrients such as zinc (Zn), iron (Fe), copper (Cu), manganese (Mn), boron (B), and sulfur (S) are also required in minimum quantity to support healthy plant growth. The recent soil health survey report (2018–2019) shows that the soil quality of the district is facing major quality challenges due to overexploitation. Detailed information on status of soil nutrients is presented in Table 1.3.

1.2.5.3 Macronutrients

Nitrogen Status

Soil health report of 2017–18 reveals that the district suffers from nitrogen deficiencies in around 73% of its soil. In fivefold classification of nitrogen availability, the zero percent share is reported under the very high and high nitrogen category. The available nitrogen (27%) shares fall under only in medium (86%) and low (14%) quality categories in the district (Fig. 1.5). Among the blocks, Harishchandrapur-I (81.2%) has the lowest nitrogen while Kaliachak-I (63.3%) has better nitrogenous soil (Table 1.3).

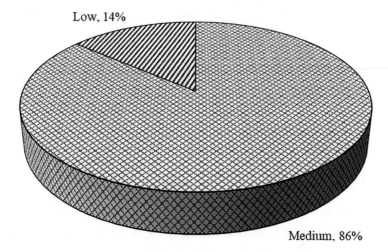

Malda District
Available Nitrogen in Soil 2018-19

Low, 14%

Medium, 86%

Fig. 1.5 Available nitrogen in soil in Malda District 2018–2019. (Source: data from Soil Health Card Report 2018–2019)

Organic Carbon

As like nitrogen, organic carbon is also an essential element of fertile soil. It enhances soil productivity. In the 2018–2019 report, around 73% of soil lacks organic carbon in the district. Even in 27% available organic carbon, more than 73% lies in low quality category, while medium and high category posits just 25% and 2%, respectively (Fig. 1.6). Eleven out 15 blocks have reported more than 70% deficiency. Again, Harishchandrapur-I (81.9%) suffered the most, and Kaliachak-I (61.5%) displays comparatively better quality among the blocks in the district (Table 1.3).

Phosphorus

Optimum share of phosphorus in soil helps to convert other nutrients for use in the building block of plant growth, while excess concentration harms. In good fertile soil, around 29% of phosphorus share an NPK ratio. Table 1.3 shows that more than 96% of soil has available phosphorus content. The low phosphorus deficiency happens mainly because of over fertilization of soil. The very high (50%) and high (21%) concentration suggest less fertility which requires nitrogen and organic carbon to compensate the excess concentration (Fig. 1.7). In six blocks, more than 99% of soil has phosphorus concentration in high or very high categories.

Malda District
Available Organic Carbon in Soil 2018-19

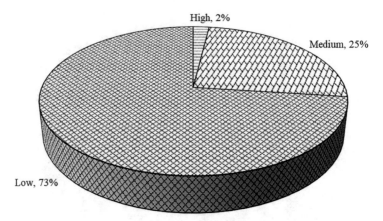

Fig. 1.6 Available organic carbon in soil in Malda District 2018–2019. (Source: data from Soil Health Card Report 2018–2019)

Malda District
Available Phosphorus in Soil 2018-19

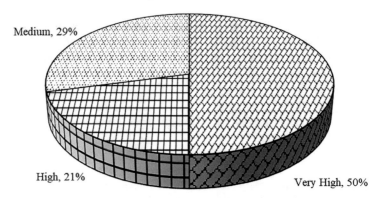

Fig. 1.7 Available phosphorus in soil in Malda District 2018–2019. (Source: data from Soil Health Card Report 2018–2019)

Potassium

In optimum NPK ratio, potassium shares around 14%. Too much concentration of potash can be detrimental to plant growth. In the district, more than 90% of soil has available potassium out of which more than 57% of soil has very high concentration, while remaining 29% and 14% of soil is reported with high and moderate potash abundance, respectively (Fig. 1.8). Among the blocks, Kaliachak records

Malda District
Available Potassium in Soil 2018-19

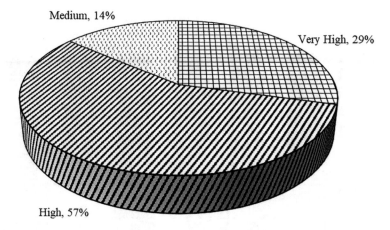

Fig. 1.8 Available potassium in soil in Malda District 2018–2019. (Source: data from Soil Health Card Report 2018–2019)

highest concentration, while Harishchandrapur-I reports lowest in the district. In ten blocks, more than 90% of soil has excess potassium concentration which is alarming (Table 1.3).

1.2.5.4   Micronutrients

Micro soil nutrients are important for the growth and development of crops. Although, it is required in low amount but still is essential for the life cycle of plant. Among major six micronutrients, the shares of zinc, boron, and sulfur are scarce while iron, copper, and manganese are recorded sufficient in the district (Fig. 1.9).

1.3   Climate

The climate of Malda District is tropical monsoon type. In general, the climate of Malda is recognized as wet and moist summer and cold dry winter. Based on weather phenomena, the district has experienced four different seasons in a year. The cold moist season starts from mid of November and continues till the end of February. During this season, the district receives low amounts of rainfall as a result of western disturbance, and average temperature of this season roars around 18 degree centigrade. Next to cold, the hot season follows from March to May with an average surging temperature of 28 degree centigrade. The hot season also receives a pre-monsoon shower along with cyclones and hailstorms. From the first week of

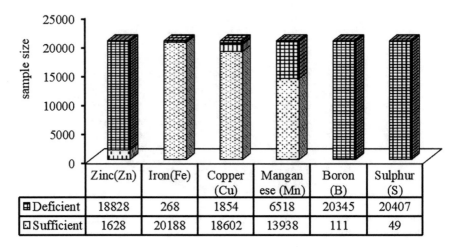

Fig. 1.9 Micro-nutrient status of soil in Malda District 2018–2019. (Source: data from Soil Health Card Report 2018–2019)

June to the end of September, the district receives more than 85% of its total rainfall from southwest monsoon which gives its name as southwest monsoon season. Followed by this, the post-monsoon season starts from October and continues up to mid of November. From here, the average temperature starts to fall and the winter season begins.

1.3.1 Temperature

The temperature of Malda District keeps on changing with the change of seasons. In normal conditions, the average lowest minimum temperature is recorded in the month of January which is considered as the coldest month. The average maximum temperature is recorded during April and May (Fig. 1.11). From August, the mean minimum and maximum temperature starts to fall, and this continues till the end of January. During this period, the daily temperature inversion is also low compared to hot and monsoon seasons. The increasing mean temperature from February to May is due to northward movement of sun, and this indicates the beginning of hot summer season in the district.

The trend of mean temperature over 30 years shows that the district is experiencing climate change phenomena (Fig. 1.10). The mean temperature graphs are showing increasing trend over the years. And this is due to global and national temperature changes (Figs. 1.11 and 1.12).

Malda District
Trends of Mean Annual Temperature

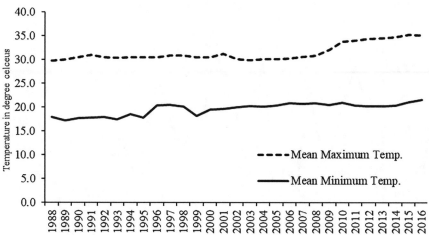

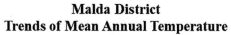

Fig. 1.10 Trends of mean annual temperature in Malda Distric. (Source: data collected from Deputy Directorate of Agriculture (Admin.), Malda 2018)

Malda District
Mean Monthly Temperature 2016

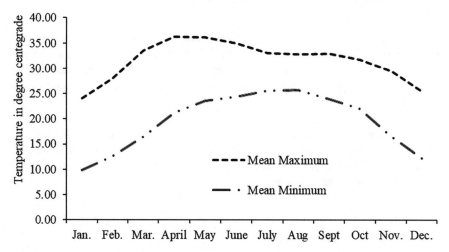

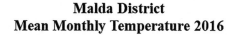

Fig. 1.11 Mean monthly temperature in Malda District 2016. (Source: data collected from Deputy Directorate of Agriculture (Admin.), Malda 2018)

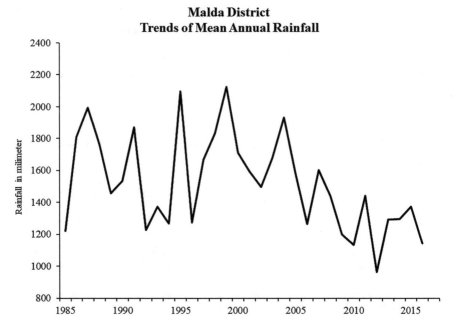

Fig. 1.12 Trends of mean annual rainfall in Malda District. (Source: data collected from Deputy Directorate of Agriculture (Admin.), Malda 2018)

1.3.2 Rainfall

In terms of rainfall, the district experiences tropical wet climates with no dry season. Figure 1.13 shows that in normal conditions there is no dry month in the district. Around 20% of total rainfall occurs during pre-monsoon, winter, and summer seasons while the remaining percentage has been recorded in May to August (monsoon season). In 2016, there are three dry months, and overall, the district receives lower monsoon rains than normal.

The yearly total normal rainfall in the district is 1485 mm (Fig. 1.13). Figure 1.12 illustrates constant variability in rainfall with decreasing trends from 1985 to 2016. From 2004 onward, the district has received constantly lower rainfall than normal and previous years' mean rainfall.

1.3.3 Relative Humidity

The percentage of water vapor present in the atmosphere at a given temperature is called humidity. Humidity plays an important role in local weather conditions. The relative humidity which is the ratio of actual amount of water vapor present in the

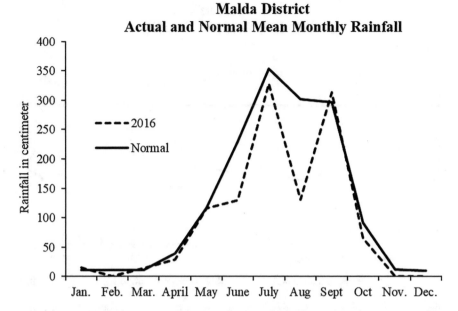

Fig. 1.13 Mean monthly actual and normal rainfall in Malda District 2016. (Source: data collected from Deputy Directorate of Agriculture (Admin.), Malda 2018)

atmosphere to maximum holding capacity remains higher compared to absolute and actual humidity. During the monsoon season, it goes up, while in the hot summer, it falls. The mean minimum and maximum relative humidity suggests an increasing trend from 1995 to 2005 which starts decreasing thereafter. From 2015 onward, the percentage of relative humidity is decreased, but average minimum and maximum humidity gap bit increased. Although relative humidity graph remains almost constant over two decades but the variability in rainfall is increased with mostly in decreasing trend (Fig. 1.14).

1.3.4 Special Weather Phenomena

Every year during the month of April and May, the cyclone and depression that originate in the Bay of Bengal also affect the district. As a result, the district receives short, heavy, and torrential rain with hail- and thunderstorms. This phenomenon is locally known as "Kalbaisakhi." This pre-monsoon shower sometimes causes huge damage to horticulture and cereal crops. In December and January, the cold winter along with occasional fog also affects Rabi crops badly.

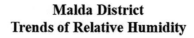

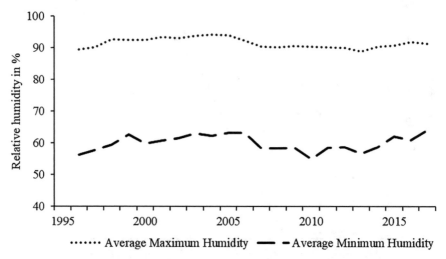

Fig. 1.14 Trends of relative humidity in Malda District. (Source: data collected from Deputy Directorate of Agriculture (Admin.), Malda 2018)

1.3.5 Weather Aberration

The district is exposed to natural calamities like floods, drought, and hailstorm very often. The Barind region of the district, because of its high altitude, generally is not affected from flood, but drought conditions prevailed several times in the recent past. The pre-monsoon hailstorm also affects this region, and resultantly heavy losses of life and property are recorded from time to time.

Along with Tal and Diara, some parts of the Barind region have experienced heavy damage due to recurrence of flood. From 1950, the district came under heavy inundation in 1971, 1987, 1991, 1999, 2001, and 2017. In the 1978 and 1996 floods, Tal and Diara region were completely washed away. Each year Malda District is exposed to moderate to severe flood because of locational channel of river Ganga on southwest side and Mahananda river in midway of the district. The economic and property losses by flood are huge, and therefore sometimes Malda's flood is announced as a national disaster.

Flood in Malda is caused by two different reasons; first, if the district receives excess rainfall then Tal and Diara regions drain more water due lower altitude from mean sea level. Figure 1.15 reports that before 2004, more than 14 times in 19 years,

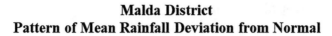

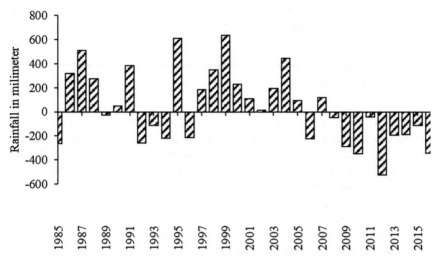

Fig. 1.15 Pattern of mean rainfall deviation from normal in Malda District. (Source: data collected from Deputy Directorate of Agriculture (Admin.), Malda 2018)

the district had experienced flood-like situation. The second reason associated with occurrence of flood in the district is heavy inundation of other regions from where river Ganga and Mahananda receive water into their channel.

1.4 Flora and Fauna

Tal and Barind region of the district are bestowed with tropical evergreen and tropical deciduous trees. Tropical deciduous forest, shrubs, jungles, and bushes are found in the Barind region also. The tree species like babul, pakur, neem, bat, semul, bamboo, jackfruits, peepal, and mango trees have grown, and these are found almost in every region of the district. In Malda District, there are only two known reserved forest areas in Adina and Gour.

Due to increasing population, some of the public forest areas have been cleared, and those lands are either converted into agriculture fields or used for settlement. Some animal species have become extinct and endangered as forest and jungle have been cleared. Presently, there are different varieties of fishes breeding in wetlands (ponds, river, beel, and lakes) mainly for local consumption and limited export.

1.5   Demographic Profile

In this section, the population distribution, density, age-sex structure, and literacy have been discussed through data and figures.

1.5.1   Distribution

Malda District shares 4.7% area and holds 4.1% of state population (Statistical Abstract, W.B, 2017). The total population of the district was 3,988,845 in the 2011 Census. This size of population resides in 15 blocks. The Kaliachak-I block is the home of 3,925,517 population which is 9.84%, the maximum in total district share, while Bamangola shares just 3.61% of the district population. In terms of area and population, Gazole is the biggest and Kaliachak-I stood as the largest block in the district. And on the same line, Kaliachak-I is the smallest, while Bamangola is the least populated block (Table 1.4). The Tal region has a maximum population (47.57%), and the Barind region has the least (21.43%) among the three regions of the district.

The district homes just 13.58% of urban population. In the last five census decades, the district has added 10.4% of urban dwellers indicating a slow pace of urbanization. With 86.42% rural population, the district witnesses negative female sex ratio, i.e., 944. And child (0–6 years of age) sex ratio stood as 950 girls per 1000 boys.

Table 1.4 Population distribution, density, and share of Malda District 2011

Block	Area (sq. km)	District %	Population	District %	Density
Harishchandrapur-I	171.41	4.6	199,493	5	1164
Harishchandrapur-II	217.21	5.8	251,345	6.3	1157
Chanchal-I	162.14	4.3	204,740	5.13	1263
Chanchal-II	205.22	5.5	205,333	5.15	1001
Ratua-I	230.53	6.2	275,388	6.9	1195
Ratua-II	173.93	4.7	202,080	5.07	1162
Gazole	513.65	13.8	343,830	8.62	669
Bamangola	205.91	5.5	143,906	3.61	699
Habibpur	396.07	10.6	210,699	5.28	532
Old Malda	215.66	5.8	156,365	3.92	725
English Bazar	251.52	6.7	274,627	6.89	1092
Manikchak	321.77	8.6	269,813	6.76	839
Kaliachak-I	105.37	2.8	392,517	9.84	3725
Kaliachak-II	222.73	6.0	210,105	5.27	943
Kaliachak-III	260.12	7.0	359,071	9	1380
District	3733	100.0	3,988,845	100	1069

Source: District Statistical Handbook, Malda 2015

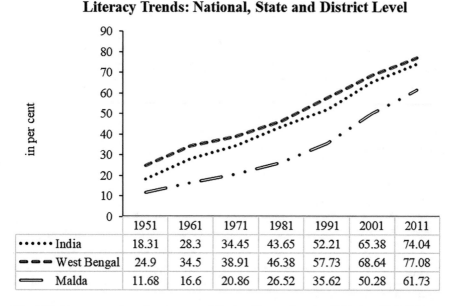

Fig. 1.16 Literacy trends: national, state and district. (Source: based on PCA data, W.B. 2011)

1.5.2 Density

The number of people per 1 km² area stood as population density. In the 2011 Census, the district population density was 1069 people per square kilometer, up from 881 in 2001. The most density is found in Kaliachak-I block (3725) and lowest density block is Habibpur (532). The higher population density than the district average is reported in 8 out of 15 blocks in 2011.

1.5.3 Literacy

A person aged 7 years and above; who can read and write with understanding in any language shall be identified as literate (Census of India). According to this definition, the literacy of the district stood 61.73% in which male literacy is 66.24% while female literacy is 56.96%. The literacy trend of the last seven censuses (from 1901 to 2011) indicates a clear male-female disparity (Fig. 1.16). Table 1.5 demonstrates the spatial pattern literacy in which Bamangola block (68.09%) and Harishchandrapur-I (52.47%) stood as the most and least literate block in the district.

The highest male and female literacy, is reported from Bamangola (75.52%) and Kaliachak-I (62.25%) blocks in the district. On the other hand, Harishchandrapur-II (57.21) and Harishchandrapur-I (47.21%) blocks have noted the least male and

Table 1.5 Literacy in Malda District 2011

Blocks	Male	Female	Total
Bamangola	75.52	60.20	68.09
Chanchal-I	68.76	61.22	65.09
Chanchal-II	59.97	54.66	57.38
English Bazar	66.96	58.88	63.03
Gazole	69.79	56.13	63.07
Habibpur	66.69	50.74	58.81
Harishchandrapur-I	57.37	47.21	52.47
Harishchandrapur-II	57.21	51.23	54.34
Kaliachak-I	68.13	62.25	65.25
Kaliachak-II	69.60	59.93	64.89
Kaliachak-III	59.91	48.07	54.16
Manikchak	64.18	50.89	57.77
Old Malda	65.25	53.66	59.61
Ratua-I	64.17	55.81	60.13
Ratua-II	58.31	53.98	56.19
District	**66.24**	**56.96**	**61.73**

Source: District Statistical Handbook, Malda 2015

female literacy respectively. All these blocks lag behind state (77.08%) and national (72.04%) average literacy (Fig. 1.17). The male-female literacy gap persists from 1951 to 2011, but wide gap started to shrink from 2001 (Fig. 1.17). Table 1.5 reveals

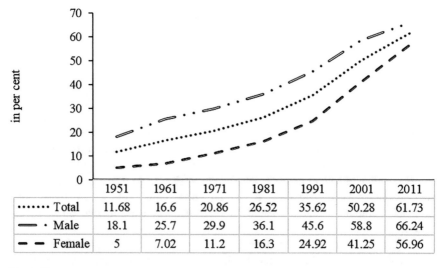

**Malda District
Litercay Trends: Total, Male and Female**

	1951	1961	1971	1981	1991	2001	2011
······· Total	11.68	16.6	20.86	26.52	35.62	50.28	61.73
──· Male	18.1	25.7	29.9	36.1	45.6	58.8	66.24
── Female	5	7.02	11.2	16.3	24.92	41.25	56.96

Fig. 1.17 Literacy trends: total, male and female in Malda District. (Source: based on census data)

that the maximum and minimum literacy gap in 2011 is found in Habibpur (15.95%) and Ratua-II (4.33%). Gender gap in literacy is higher in eight blocks than the average gap in district (9.28%). Despite the dedicated program for educational upliftment by state and central governments, the literacy in Malda district is comparatively much lower than national and state average (Rahaman and Rahaman 2018).

1.5.4 Age-Sex Compositions

Population characteristic is an important determinant in planning and development of a region. The issues like age-specific workforces, gender-wise population characteristics, and gender-specific planning are very much involved with age-sex composition of population of a region. Figure 1.18 reports the age-sex pyramid of population for Malda District. It locates about 57.87% of population of the district in the working age group (15–59 years). In the same age group population, the male population accounts 58.29% and females 57.41%. In the aged population group (>60+), shares of female surpass male, and the same is noted in case of below 15 age groups. In the overall population pyramid (Fig. 1.18), young (< 18 years) and adult (15–30 years) population share 40.94% and 46.82%, respectively.

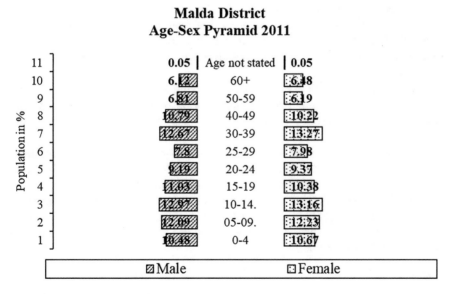

Fig. 1.18 Age-sex pyramid of Malda District 2011. (Source: based on author calculation data from statistical handbook of Malda 2015)

1.6   Economic Aspects

The activities that generate income and provide livelihood options are considered as economic activities. In this section, economic activities like agriculture, industry, market, nature of workforce, per capita income, and transport have been discussed in detail.

1.6.1   Agriculture

Agriculture is the mainstay of income and livelihood to more than 96% of rural population of the district (District Statistical Handbook 2015). The fertile alluvial soil along with natural irrigation sources of Indo-Gangetic plain provides ample opportunities to cultivate different types of crops in the district. The farmers of the district grow crops in three seasons (Rabi, Kharif, and Zaid). Most of the effort for cultivation is done by family labors which is also a source for crop rotation and diversification. The paddy being the dominant crop grows twice in a year. In terms of area and production, paddy, wheat, and maize are major cereal grown every year. Among vegetables, potato is the number one crop in terms of both area and production. Followed by potato, brinjal, cauliflower, cabbage, tomato, green chilies, and spices are cultivated for household's consumption and for earning as well. Among cash crops, jute, sericulture, makhana and mesta are grown in different seasons. Most households engage in horticulture and vegetable cultivation for income. The famous Malda mango has already earned fame in the international market that boosts income of middleman merchants. In recent years, litchi is grown by farmers for export purposes.

Along with cultivation of crops, farmers also domesticate animals for field cultivation and household purposes. Livestock rearing is common in the district. The wetlands are used for pisciculture which is the good source of income to few households.

The maximum number of farmers in the district falls in the marginal and small land size category, and their average size of landholding is just 0.72 ha. The marginal and small land size farmers comprise more than 95% and share more than 85% of cultivable land; therefore, corporate and commercial farming has not made any significant dent yet here (Figs. 1.19 and 1.20). Due to small land size and land parcels, farmers in the district have been practicing subsistence agriculture for the sake of family consumption.

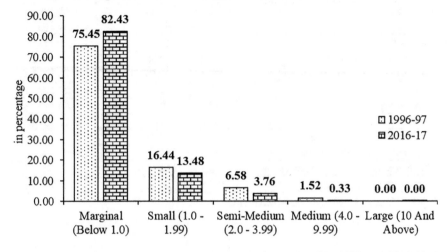

Fig. 1.19 Percentage of land size holders in Malda District in 1996–1997 and 2016–2017. (Source: based on Input Survey data, Agriculture Census 2016–2017)

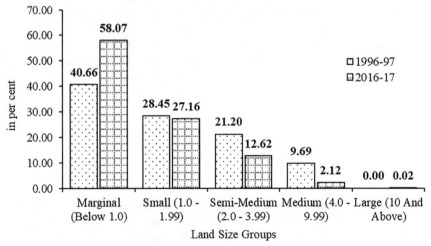

Fig. 1.20 Percentage of area under different land size classes in Malda District in 1996–1997 and 2016–2017. (Source: based on Input Survey data, Agriculture Census 2016–2017)

Table 1.6 Profession in different categories in Malda District 2011

Description	Population	% of population	Male	Female
Total worker	1,537,847	38.55	70.65	29.35
Non-worker	2,450,998	61.45	39.37	60.63
Main worker	1,050,995	14.26	81.06	18.94
Main worker – cultivator	219,241	20.86	94.07	5.93
Main worker – agricultural laborers	322,452	30.68	87.16	12.84
Main worker – household industries	98,383	9.36	25.29	74.71
Main worker – others	410,919	39.10	82.69	17.31
Marginal worker	486,852	31.66	48.17	51.83
Marginal worker – cultivator	35,941	7.38	55.37	44.63
Marginal worker – agricultural laborers	223,307	45.87	57.45	42.55
Marginal worker – household industries	98,304	20.19	11.23	88.77
Marginal workers – others	129,300	26.56	58.24	41.76

Source: Primary Census Abstract, Malda District 2011

1.6.2   Industry

The district does not have any known heavy industry. The main hurdle to industrial development in the district remains lack of known mineral resources, dedicated infrastructures, and planned government policies (Hussain 2011). A few local household industries like rice mills, food processing industries, and jute processing farms are there but they have been facing institutional and infrastructural constraints. And therefore, working conditions at household industries (bidi workers, pot makers, and bakery food workers) are seasonal and hence they earn some amount from outsourcing or as a short term migrant labour. Though the district has enough scope to develop mango processing industries, jute mills and food parks based on local raw materials but encouraging government policy is still far reaching (Das 2017).

1.6.3   Workforce

In 2011 Census, the district counts total workers and non-workers as 1,537,847 and 2,450,998, respectively (Table 1.6). Hence, the work participation rate stood 38.55% in which the male number, 1,086,461 (72.7%), dominates female, 451,386 (28.3%). The main workforce, who works for more than 180 days in a year, has shared just 14. 26% in which again male workers are more than 81%. In the marginal (work less than 180 days in a year) workforce category, the common characteristic evolves out

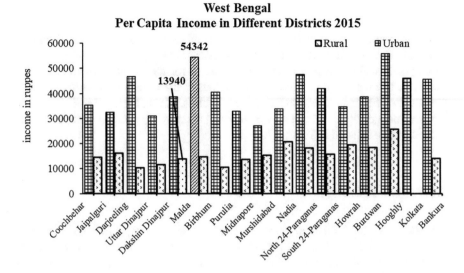

Fig. 1.21 Per capita income in different districts of West Bengal 2015. (Source: based on State Statistical Handbook 2015)

as the share of male dominating over female workforce, and same remains true in the case of main working class. In total cultivators including main and marginal are just 16.59% because of limited cultivable land abridged by overpopulation. In total workers, agricultural laborers share 35.48%, while household industrial workers and other workforces share 12.79% and 35.12%.

1.6.4 Income

The total earning from all different sources of a unit of area per say state or district is called gross income. After deduction of the cost, the remaining sums as total income. One the other hand, the per capita income (PCI) is extracted after dividing the total income by the total number of populations. The per capita income is an important indicator of an economy. The higher PCI indicates that the source of earning is either from secondary or tertiary economic activities, and low PCI suggests an agriculture-based economy.

Figure 1.21 shows lower PCI in case of rural areas than urban. The highest rural per capita income is for Hooghly (25,635 rupees) and lowest counts for Purulia (10,677 rupees) in the state of West Bengal. In case of rural PCI, Malda rank ninth, but it is second for urban PCI in the state. The reason for low rural income is due to an agricultural base. TThe disguised unemployment in agricultural sector has remained the main cause of low per capita income in rural areas of the district.

1.7   Market

Available market infrastructure is an important indicator to agriculture development of a region. As of December 31, 2015, the district has 2 principal markets and 24 sub-markets which are selling and buying place for almost all agricultural commodities. The nature and periodicity of markets in the district vary from regulated to open, and big markets mostly sit a weekly basis. Malda District remains important for mango, litchi, jute, paddy, and fish marketing. A detailed account of the market has been listed in Table 1.7.

1.7.1   Cold Storage

Till March 30, 2018, there are only two public-run cold storages in the district. The cold storage at Samsi (Ratua-I block) with 4000 million tons storing capacity for potato and in Old Malda (not fully functional yet) are in the list. However, the capacity of these cold storages might be expanded up to 10,000 million tons. Two new multipurpose cold storages have come up in the private sector under "Credit Linked Capital Subsidy Scheme." The "Pack House of Malda" also has a cold storage of 240 million.

A detail account of different markets is depicted in Table 1.7. The nature of markets remains mostly weekly in nature and covers up to 65 km from site. In terms of accessibility, Old Malda block has four markets, while Harishchandrapur-II, Gazole, Habibpur, and Manikchak blocks get access of two markets each in a week. The items like cereals, vegetables, and livestocks have been bought and sold in these markets throughout the years. During June and July; the supply of mango, litchi, and other seasonal fruits has increased. Jute, makhana, and other non-food crops are found seasonally in selected market of the district. The nature and type of transportations determine the areal coverage and number of people to be served by a market.

1.8   Transport

The means and medium of transportation are the lifeline of an economy. Growth and development of an economy is interlinked with transportation. The district is well connected through national highways (NH) and railways with the rest of India. The NH 34 passes over Malda, connecting two major cities of West Bengal, namely, Kolkata and Siliguri. The newly constructed NH 81 links Harishchandrapur with Gazole to Chanchal and Samsi; by doing this, access of weekly and cooperative markets became fast and easy. These well-connected road networks help inter-district and interstate travel of goods and passengers. The interstate transport makes more convenient by well-connected railway, which passes over almost the middle of

Table 1.7 Available marketing option in Malda District 2015

Block	Name of the market center	Periodicity	Important commodities handled	Area covered	No. of farm families covered
Harishchandrapur-I	Kushida Hat	Weekly	Cattle, paddy, wheat	5–10 km	1600
Harishchandrapur-II	Talgram Hat	Weekly	Paddy, wheat, vegetables, cattle	2–15 km	1800
	Tulshihata				
Chanchal-I	Chanchal Hat	Weekly	Paddy, wheat, vegetables, cattle	5–15 km	2100
Chanchal-II	Maoltipur Hat	Weakly	Vegetables, paddy, wheat, pulses	2–20 km	750
Ratua-I	Samsi Regulated Market and Balupur Hat	Weekly	Oilseeds and vegetables	2–50 km	3500
Ratua-II	–	–	–	–	–
Gazole	Gazole Hat	Weekly	Mustard, paddy, vegetables	5–20 km	1250
	Alampur Hat				
Habibpur	Aiho Hat	Weekly	Cocoon, paddy, wheat, vegetables	5–25 km	850
	Bulbulchandi Hat				
Bamangola	Pakuaa Hat	Weekly	Paddy, wheat, vegetables	5–25 km	1300
Old Malda	Nawabganj Hat	All weekly	Jute, wheat, goat, cattle	5–15 km	1300
	8 Mile Hat				
	Ballatali Hat				
	Sanibari Hat				
English Bazar	English Bazar Regulated Market	Daily	Mango	1–65 km	3000
Manikchak	Mathurapur Hat	Weekly	Paddy, wheat, vegetables	1–15 km	500
	Champanagar Hat				
Kaliachak-I	Babur Hat	Weekly	Cocoon, paddy, wheat	1–25 km	1100
Kaliachak-II	Gosai Hat	Weekly	Rice	1–15 km	600
Kaliachak-III	Sahabajpur and Golapganj Hat	Weekly	Cattle	1–30 km	500

Source: CDAP XII plan Malda 2015. Another multipurpose cold storage is coming up at Adina in the cooperative sector with a proposed capacity of 10,000 million tons. With this, the scope of storing seasonal fruits, vegetables, and other perishable commodities has improved tremendously over the last two decades. As regards of godowns (storehouse), 34 rural godowns are in the various stages of completion in the district under "Credit Linked Capital Subsidy Scheme." For financing rural godown, subsidy is being routed through National Bank for Agriculture and Rural Development (NABARD). The total capacity generated is to the tune of 30,500 million tons. In addition, "Cooperative Agriculture and Rural Development Bank Ltd" of Malda has financed nine rural godowns for a total capacity of 5139 million tons

Table 1.8 Trends of road length in Malda District 2014

Year	Surface road	Unsurfaced road	Total length
2009–2010	2981	4173	7154
2010–2011	3134	3853	6987
2011–2012	3674	4528	8202
2012–2013	3771	4517	8288
2013–2014	3897	4453	8349
2014–2015	4254	3423	7678

Source: District Statistical Handbook 2015

Table 1.9 Transportation infrastructure in Malda District 2015

Name of block	Surface road	Unsurfaced road	No. of ferry service	No. of bus route	Distance of nearest Railway Station from Block HQ (in KM)
Harishchandrapur-I	270.33	366.9	–	3	2
Harishchandrapur-II	226.96	362.35	7	2	5
Chanchal-I	262.45	214.65	4	4	13
Chanchal-II	275.11	274.85	3	1	6
Ratua-I	215.94	440	6	5	13
Ratua-II	385.78	412.83	3	3	12
Gazole	309.59	106.16	8	9	2
Bamangola	267.17	443.87	14	2	22
Habibpur	287.25	298.23	4	4	6
Old Malda	160.54	164.7	1	4	2
English Bazar	236.2	92.7	–	7	2
Manikchak	193.26	384.45	10	1	5
Kaliachak-I	246.7	374.69	2	3	3
Kaliachak-II	204.97	230.95	–	5	8
Kaliachak-III	270.73	278.29	2	4	5
District	3250.67	4445.62	64	57	–

Source: Statistical Handbook, Malda District 2015–2016

the district. Except Habibpur, Bamangola, and Manikchak; remaining 12 blocks are well connected through railway and offer service for transportation of passengers and goods (Table 1.9).

The impact of transportation on local economy is enormous. And, the true realization of this is anticipated through state-sponsored roadways. The Pradhan Mantri Gramin Sadak Yojna (PMGSY) implements to construction metaled surface roadways. Under PMGSY scheme, the length of roadway has increased in the district (Table 1.8). The local governing bodies (panchayat samiti and zila parshad) encourage to construct metaled roads in the district.

References

Das, S. (2017). *Changing hydro geomorphological characteristics of Kalindri river Malda District and their impacts on riparian livelihoods a geographical overview.* Retrieved from https://shodhganga.inflibnet.ac.in/handle/10603/215373: http://hdl.handle.net/10603/215373

Rahaman, M., & Rahaman, H. (2018). Gender disparity in literacy in Malda District. *International Journal of Research in Social Sciences, 8*(10), 123–141.

Chapter 2
Land Use and Cropping Pattern Dynamic

In the previous chapter, a detail analysis of study area has been describedwith recent data. Physical and human aspects, especially related to agriculture, are explained in light of the changing agricultural scenario of the district. The present chapter is committed to understand and explain one of the aspects of changing agriculture, i.e., "land use and cropping pattern dynamics". In this chapter, the conceptual, theoretical, and practical aspects of land use and cropping patterns with respect to time and space have been described. To assess the crop diversification of any region, it is necessary to understand the extent of cropping pattern and change in it. The theoretical and conceptual aspects of land, land use, and types of land use are explained with national and international perspectives. The practical aspect of land use and cropping pattern dynamics focuses on changing agricultural land use and cropping pattern by land size classes at block level from 1995–1996 to 2015–2016. This follows the assessment of the amount of change in cropping pattern across land size classes. The last section deals with nature and reasons for changing cropping patterns from field survey data.

2.1 The Land

The land is a stage where all human activity is performed. Human activity on land gives sense to "land use" which refers to the purpose it serves. The varied needs of land from food production to shelter, recreation, extraction, and so on are being shaped by two border forces including human necessity and natural features and process (Briassoulis 2019, 3). The outcome of change of land use is sometimes beneficial and at times harmful as the process requires human intervention. The positive effect of human and environmental processes on land has a long scientific and lay history especially where people used land as a resource. From ancient writing to present new millennium land use documents, the science, literature, and

H. Rahaman, *Diversified Cropping Pattern and Agricultural Development*,
https://doi.org/10.1007/978-3-030-55728-7_2

folklore overwhelmingly focus on complex land use dynamics. In present study, the varied land use changes with time, and geographic space is being studied in light of agricultural land use concerning description and explanation of pattern of change.

The study of land use is time and space dependent. It is therefore required to look at some frequent and official sources concerning land use.

The Food and Agriculture Organization (FAO) defines land as "an area of the Earth's surface" (1996). This definition is refined and holistic as it includes past, present, and future of all land use dimensions encompassing hydrosphere, biospheres, and lithospheres. In scholarly viewpoints, the definition of land differs because different priorities were given to different attributes and characteristics of land from a natural and social sciences perspective. Wolman (1987) defines land from a natural science viewpoint and so does FAO. On the other hand, economic attributes got priority in the definition of Hoover and Giarratani (1984, 1999). They emphasize that land "first and foremost denotes space.... The qualities of land include, in addition, such attributes as the topographic, structural, agricultural and mineral properties of the site; the climate; the availability of clean air and water; and finally, a host of immediate environmental characteristics such as quiet, privacy, aesthetic appearance, and so on."

Available literature suggests that land as a natural entity occupies certain theoretical space in individual imagination and therefore "the land" is a multidimensional and dynamic concept. The experience of land exhibits varied meaning to certain space and scholars; hence the idea of land use too differs at spatial and temporal context.

2.2 The Concept of Land Use

The term "land use" is sometimes synonymously used with the word land covers. At international level, "key sources of global data do not distinguish clearly between land cover and land use" (Meyer and Turner 1994). The importance of necessary distinction between "land use" and "land cover" is stated by Graetz (1994) who emphasizes on "to retain the definition of grassland by ecological attributes (vegetation structure and composition) rather than by its principal use, livestock production. ... it is not possible directly to relate land use as such to the major physical processes of global environmental change. Land use cannot be directly related to these forms of global change because it is a qualitative descriptor. Land use categories are abstract typologies that, although useful, cannot be meaningfully included in process models seeking to forecast the time and space patterns of global change. It is land cover, rather than land use, that has the mechanistic meaning in the processes of global environmental change." Ideally, literature on land use and land cover is clearly distinct in spatial and physical context. The physical state of land includes cropland, mountain, forest, water, and other earth materials. Turner II et al. (1995a, b) note "land cover is the biophysical state of the earth's surface and immediate subsurface. Moser (1996) further adds that "the term land cover originally referred to the types of vegetation that covered the land surface, but has broadened subsequently to include human structures, such as buildings or payment, and other aspects of the physical environment, such as soils, biodiversity, and surfaces and groundwater."

Turner II et al. (1995a, b) brief "land use involves both the manner in which the biophysical attributes of the land are manipulated and the intent underlying that manipulation – the purpose for which the land is used." The international organizations like FAO (1995) added, "land use concerns the function or purpose for which the land is used by the local human population and can be defined as the human activities which are directly related to the land, making use of its resources or having an impact on them." In academic discussion, study of land use is context driven such as territorial scales, urban scale, agricultural scale, and so on. According to Chapin and Kaiser (1995), "At territorial scales involving large land areas, there is a strong predisposition to think of land in terms of yields of raw materials required to sustain people and their activities. At these scales, 'land' is a resource and 'land use' means 'resource use'. In contrast, at the urban scale, instead of characterizing land in terms of the production potential of its soils and its sub mineral content, the emphasis is more on the use potential of the land's surface for the location of various activities. In spatial and temporal context, the description of land use includes mixed types of form, pattern, extent, intensity, tenurial status of land" (Bourne 1982).

A clear distinction between land use and land cover can be understood from subtypes classification. For example, agricultural land is a type of land cover, and within it cropland, orchard, recreation, mixed use, and groves are types of land use. Meyer and Turner (1994) interpret, "By land cover is meant the physical, chemical, or biological categorization of the terrestrial surface, e.g. grassland, forest, or concrete, whereas land use refers to the human purposes that are associated with that cover, e.g. raising cattle, recreation, or urban living." And sometimes both land use and land cover are understood as the same. As Turner and Meyer (1994) explain, "A single land use may correspond fairly well to a single land cover: pastoralism to unimproved grassland, for example. On the other hand, a single class of cover may support multiple uses (forest used for combinations of timbering, slash-and-burn agriculture, hunting/gathering, fuelwood collection, recreation, wildlife preserve, and watershed and soil protection), and a single system of use may involve the maintenance of several distinct covers (as certain farming systems combine cultivate land, woodlots, improved pasture, and settlements). Land use change is likely to cause land cover change, but land cover may change even if the land use remains unaltered."

2.3   The Land Use Dynamic

Land use dynamics simply mean a change (positive or negative) of areal extent of a particular type of land at a point of time in a region. Dynamic of land is measured quantitatively. Depending upon scale, extent, and level of measurement, the nature of change is recorded. At global level, land use change detected at larger scale aims at major areal details. In broader sense, land use change confuses with conversion and modification (Turner II et al. 1995a, b). Land conversion refers to the change in mix and pattern of land in a region, while modification denotes alteration of land structure such as productivity or price of land from low to high or vice versa. More specifically, Meyer and Turner (1996) differentiate: "Land use (both deliberately and inadvertently) alters land cover in three ways: converting the land cover, or changing it to a qualitatively different state; modifying it, or quantitatively changing its condition without full conversion; and maintaining it in its condition against natural agents of change." Jones and Clark (1997) state agricultural land use change as "intensification, intensification, marginalization and abandonment."

Land use change is a continuous process caused by many interrelated drivers such as biophysical and socioeconomic. The biophysical drivers comprise all those characteristics and processes that drive in natural environments like landform and topography, geology and soil, weather and climate, drainage, natural vegetation and plan succession, geomorphic processes, and volcanic eruptions (Briassoulis 2019, 13). The socioeconomic drivers include economic, political, social, demographic, and institutional factors and population change, technological change, industrial change, and change in family, market, government policy, and processes (Briassoulis 2019, 13).

There are various factors and processes contributing together or in simultaneous way which can be categorized into three distinct driving forces: human driving forces, human mitigation forces, and proximate driving forces. The human driving forces or "macro forces are those fundamental societal forces that in a causal sense link human to nature and which bring about global environmental changes" (Moser 1996, 244). The socioeconomic organizations (markets, and political institutions, ecology, economy) and technological and demographic and population change are the examples of micro forces of land use change. The human mitigation forces comprise "those forces that impede, alter or counteract human driving forces" (Moser 1996, 244). The prominent example of mitigation forces are local, regional, and international rules and regulations, technological innovations, informal social regulations, and market adjustments (Briassoulis 2019, 15). The proximate driving sources "are the aggregate final activities that result from the interplay of human driving and mitigating forces to directly cause environmental transformations, either through the use of natural resources (e.g. as input to agriculture, mining activities, or as raw materials for industrial production), through the use of space, through the output of waste (solid waste, emissions, pollution, etc.) or through the output of products that in themselves affect the environment (e.g. cars, plastic bags)" (Moser 1996, 244–245 cited by Briassoulis 2019, 15–16).

The land use discourse focusing on man's action is an epitome of change, modification, use and reuse, dispersion, and evacuation of land. Hence, in land use dynamics, man is the master of possible use for house, roads, agriculture, and others. Meyer (1995) argues "land use is the way in which, and the purpose for which, human beings employ the land and its resources." A study done by Turner et al. (1994)

concludes that land use is nothing but human employment of land for appropriate production processes as determined by complex socioeconomic factors.

The continuous and untiring interventions of man are the driving force in cross-disciplinary analyses. With time and space, all human activities are shaped and determined by varying processes of land. The land use unit and land use system are such outcomes of varying processes. The land unit includes soil, topography, and vegetation sites, while the land system implies the concept of assembled land units which are geographically and genetically related and constitutes a recognizable and recurring pattern. Due to the continuous intervention by nature and man on land, the land unit has been continuously modified and changed which results in new land as land use (Christian 1957). Therefore, land use and change therein is an outcome of an interplay of nature and man with time. A land system is bestowed with many resources from minerals to biological; therefore, proper management practice for future generation is the actual meaning of land use (Memoria 1984). The different use of a land unit at any given time and place is based on the infraction of social, economic, political, technological, and environmental aspects (Young 1975).

2.4 Types of Land Use

The study of land use dynamics became more diverse and specialized as well especially in the age of technological revolution (Wolman 1987). The specialized classification of land use evolves as agricultural, industrial, urban, sectoral, and land use. The agricultural land use classification primarily though follows some considered base as it was done by FAO (1950) under four categories: arable land (or cropland), grassland (or permanent pastures), forest land (or forest and woodland), and others. Remote sensing and GIS tools make it possible to classify other types of land use into more categories such as tundra land, desert land, and unmanaged rangeland. And similar micro units' types of classification ensue in other land use aspects of cities, towns, transport, etc.

In 1995, FAO revised its previous land use classification and developed a more elaborative international framework in which three hierarchical systems were developed. The second level includes a functional land use category in which 12 categories are there. At sub-global level, most of the national classifications follow purpose and context as criteria. Based on remotely sensed data, US Geological Survey developed two-level hierarchy in which first level includes 9 types and second level extends up to 36 (Kleckner 1981) extended up to substate regional level. In Canada, the Canada Land Inventory initiated land use classification (Pierce and Thie 1981). In Europe, among several land use classification systems, mainly two CORINE (Coordinated Information on the European Environment) and CLUSTERS (Classification for Land Use Statistics: Eurostat Remote Sensing Programme) were set up to prepare land use types based on hierarchical system (Beale 1997).

In India, first ever land use classification was done by a coordination committee on Agricultural Statistics in 1948 under the Ministry of Food and Agriculture

(Mohammad 1992). However, scholarly writing on land use and other aspects was accepted from the mid-1940s. Apart from dedicated classification, different facets of land use study have been enriched by S.P. Chatterjee in 1945, M. Shafi in 1951, M Hussain in 1971, and others (Mohammad 1992). The first Agriculture Census of India has considered fivefold land use classification in 1970–1971. They are:

1. Forest
2. Area not available for cultivation
3. Other uncultivated land excluding current fallow
4. Fallow land
5. Net sown area

To compete global land use classifications of FAO and to meet the requirements of India's land use planning, the Indian Council of Agricultural Research (1992) has revised land use types into nine categories which are adopted by the Agriculture Census, Government of India. These are following:

1. Forest
2. Land put to non-agricultural use
3. Barren and uncultivated land
4. Permanent pastures and other grazing land
5. Land under miscellaneous tree crops
6. Cultivable waste lands
7. Current fallow land
8. Other fallow land
9. Net sown area

After going through different dimensions of land use classification, the present study has carefully adopted classification of Agricultural Census of India. In Agriculture Census classification, the different activities on land have been the major criteria of land use classification. Therefore, Census classification is more meaningful in the present analysis.

2.5   Types of Agricultural Land Use in Malda District

In India, most of the state governments follow ninefold land use classification for planning purposes in line with the Agriculture Census. The land use classifications of the Agriculture Census have three major groups comprising the uncultivated area, the area under current fallow, and net sown area. The fallow and uncultivated lands are further classified into different categories. In brief, Agriculture Census provides data on "ninefold" land use classification which has been explained below:

Total Area (TA)
Agriculture Census defines "total area as both cultivated and uncultivated area under an operational holding. It should also comprise the land occupied by the farm

buildings including the house of the holders. The total area of an operational holding would be equal to that total geographical area for a particular household at a
point of time."

Figure 2.1 shows that the declining trend in total area is due to redefinition of
district boundary with Uttar Dinajpur and Murshidabad district in 2004. According
to the Agriculture Census, the total reported area in district was 301,797 ha in
1995–1996, and it reduced to 297,691 ha in 2015–2016. Figure 2.2 displays the
positive change in total area in Manikchak, Habibpur, Gazole, and Old Malda from
1995–1996 to 2015–2016. Other blocks record negative growth in the total area
over 20 years. The land size class wise data locates maximum increase in total area
from 120,132 ha in 1995–1996 to 171,666 ha in 2015–2016 under marginal category. The other land size groups locate negative change (Fig. 2.3). The maximum
area decreases under medium land size category (26,095 ha) followed by semi-
medium (242,988 ha), small (3958 ha), and large (1089 ha) land size in the district.

Net Sown Area (NAS)
The NSA is the actual land for cultivation for agricultural purposes in a year of a
region. The Census considers NSA as "the total cultivated area during the reference
year without any regard to a number of times it has been cultivated in a year. So, to
find out the net sown area, the area cultivated more than once during the same year
will be counted only one. The field crops and orchids both will form the part of the
net sown area."

Figure 2.1 reveals that the district experiences a declining trend in the net sown
area from 291,061 ha in 1995–1996 to 288,708 ha in 2015–2016. Some of the reasons behind this change are erratic monsoon rainfall, price instability, low profit,
and increasing cost of production (CDAP-XII 2015). The block-wise change in net
sown area reports that Gazole block gains maximum area, i.e., 8860 ha., followed

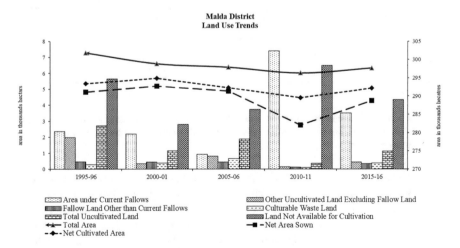

Fig. 2.1 Land use trends in Malda District 1995–1996 to 2015–2016. (Source: based on author
calculation, data extracted from Agriculture Census, Government of India)

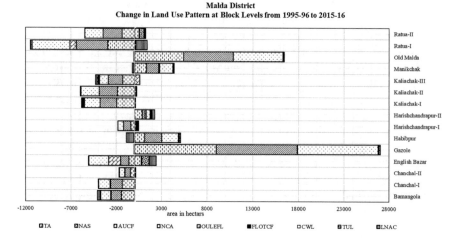

Fig. 2.2 Change in land use pattern at block levels in Malda District from 1995–1996 to 2015–2016. (Source: based on author calculation, data extracted from Agriculture Census, Government of India)

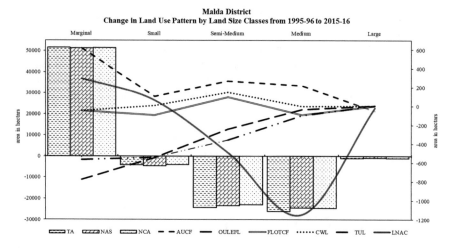

Fig. 2.3 Change in land use pattern by land size classes in Malda District from 1995–1996 to 2015–2016. (Source: based on author calculation, data extracted from Agriculture Census, Government of India)

by Old Malda (5401 ha.) and Habibpur (1902 ha.) and the remaining blocks report negative change in which Ratua-I shows negative change of 3446 ha. due to sand deposition caused by flood in 2001 and 2007 (Fig. 2.2). The land size category wise data depicts significant increases only under marginal land size class that is 51,201 ha (Fig. 2.3). The NSA in other classes displays negative change which the medium land size class records maximum decline, i.e., 24,491 ha in 2015–2016 from 1995–1996.

Area Under Current Fallows (AUCF)

The AUCF is related to "all the areas which are usually cropped but have not been cultivated during the reference year. The area to be classified as current fallow, it should be fallow during the current year and should have been cultivated during the previous years or more. If an area is not being cultivated for more than one in a year it would be categorized as old fallow or culturable waste land" (Agriculture Census). Figure 2.1 illustrates that there are increasing trends in the area under current fallow from 1995–1996 to 2015–2016 in the district. It was 2362 ha in 1995–1996 and rose to 3511 ha in 2015–2016. The reason for increasing fallow land is the decrease in net sown area.

Figure 2.2 displays the maximum increase of AUCF in Kaliachak-III (593 ha) block, and the maximum decrease is reported in English Bazar block (1335 ha) in 2015–2016. Out of 15 blocks, 5 blocks namely, Habibpur (95 ha.), Bamangola (59 ha.), Chanchal-I (53 ha.), Chanchal-II (25 ha.), Ratua-I (740 ha.), and English Bazar (1335 ha.), have recorded negative change in 2015–2016 from 1995–1996. Figure 2.3 shows that the AUCF illustrates decreases under large land size class, while other categories note increase in area. The maximum increase is found in marginal class (615 ha) followed by semi-medium (268 ha), medium (216 ha), and small (102 ha) category in 2015–2016 in district.

Net Cultivated Area (NCA)

The net cultivated area includes net sown area under current fallows with reference year. NCA is the actual cultivable land for a district or block. Figure 2.1 sits decreasing trend in the net cultivated area from 293,423 ha in 1995–1996 to 292,220 ha in 2015–2016. This negative growth is due to increased area under current fallow land.

In 2015–2016 the district added 1203 hectare area under NCA. With an increase of 8901 ha, Gazole block gains maximum NCA in district. The NCA also increased in Old Malda (5419 ha), Manikchak (1470 ha), Habibpur (1808 ha), and Harishchandrapur-II (436 ha) block from 1995–1996 to 2015–2016. The maximum area under NCA is decreased in Ratua-I (4186 ha.) followed by English Bazar (2209 ha), Ratua-II (2053 ha.), Kaliachak-II (1999 ha.), and Kaliachak-I; Chanchal-I, Bamangola, Kaliachak-III, Harishchandrapur-II, and Chanchal-II have also reported negative change in ascending order (Fig. 2.2). The marginal land size class gains 51,372 ha, while other classes have noted negative growth in NCA. Among different land size classes, the medium category exhibits maximum decline, i.e., 24,565 ha in the district.

Other Uncultivated Land Excluding Fallow Land (OULEFL)

The other uncultivated lands excluding fallow include two types of land use. For micro-level planning, the Census further classifies OULEFL into the following two categories:

Permanent Pasture and Other Grazing Lands "all grazing lands whether they are permanent pasture and meadows or not, village common grazing lands would, however, be excluded" (Agriculture Census).

Land Under Miscellaneous Tree Crops It includes "cultivable land, which is not included in the net sown area but it puts to some agricultural use. Land under casuarina trees, thatching, grasses, bamboo buses and other groves for fuel chain not included under orchards would be covered under this category" (Agriculture Census).

Figure 2.1 reveals that area under OULEFL decreased from 1987 to 428 ha in the district because built-up area has increased. The maximum increase has been noted in English Bazar (7369 ha) followed by Ratua-II (520 ha) and other blocks locating the negative change (Fig. 2.2). Except for large land class, other land size class categories have recorded negative growth. In 20 years, marginal land size class has reported the maximum decrease, i.e., 566 ha in the district, due to increasing population (Fig. 2.3).

Fallow Land Other Than Current Fallows (FLOTCF)

The FLOTCF is concerned with "all lands, which are taken up for cultivation but are temporarily out of cultivation for a period of not less than five years and more than five years. The reason for keeping such land for low cultivation is more likely poverty of the cultivator, inadequate supply of water, climate variability, river silting and others" (Agriculture Census).

Figure 2.1 depicts that the area under FLOTCF has decreased from 462 to 327 ha in the district. Figure 2.2 explains that the maximum increase in FLOTCF reports in English Bazar (82 ha) and decreases in Old Malda (74 ha). Eight out of 15 blocks sit negative changes in FLOTCE in two decades. Except the large land size class, other land classes have recorded negative growth in area under FLOTCF. The area under semi-medium category has decreased by 95 ha in two decades which is the maximum in the district (Fig. 2.3).

Culturable Waste Land (CWL)

The CWL is the social dimension of land use classification. The Agriculture Census defines "all lands available for cultivation whether not taken up for cultivation or taken up for cultivation once but not cultivated during the current year and the last five years or more in succession for one reason or the other, i.e., > 5 years in succession. Such lands may be either wholly or partly covered with shrubs and jungles, which are not put to any use. Land once cultivated but not cultivated for five years in succession would also be included in this" (Agriculture Census).

Figure 2.1 reveals an increasing trend in the area under CWL from 273 to 372 ha over 20 years in the district. The waste land has been brought back to cultivation after reclamations which is the main reason behind increase in NSA. Among the blocks, Kaliachak-II gains the maximum area in this category where Ratua-I reports negative growth of 130 ha from 1995–1996 to 2015–2016. The semi-medium group gains maximum, i.e., 147 ha, followed by all land size classes (99 ha). In the marginal land size category, the maximum decrease is 50 ha in the same two decades (Fig. 2.3).

Total Uncultivated Land (TUL)

The TUL is calculated as "the area arrived at by deducting the total cultivable area from the total reported area" (Agriculture Census). Figure 2.1 describes the decreasing trends of area under TUL due to increasing population pressure in the district. In the district, TUL was reduced from 2722 ha in 1995–1996 to 1128 ha in 2015–2016. Among blocks, English Bazar and Ratua-II have gained area under this land use category with 785 and 533 ha, respectively. The maximum decrease in area under this category is in Kaliachak-III block, i.e., 143 ha from 1995–1996 to 2015–2016 (Fig. 2.2). Except in large land size class, other land classes display a decreasing trend. The maximum decrease is observed in the marginal (777 ha) group followed by the small (545 ha) category from 1995–1996 to 2015–2016 (Fig. 2.3).

Land Not Available for Cultivation (LNAC)

The LNAC is further classified into forest area and area under non-agricultural use.

Forest This type of classification is done for environmental and ecological planning of a region. It includes "all land classified either as forest under any legal enactment, or administered as forest, whether State-owned or private, and whether wooded or maintained as potential forest land. The area of crops raised in the forest and grazing lands or areas open for grazing within the forests remain included under the 'forest area.' The only private forest is covered for the purpose of the Agricultural Census" (Agriculture Census).

Area Under Non-agricultural Use Agriculture Census takes "all land occupied by buildings or ponds and put to use other than agriculture will be included in this category."

From 1995–1996 to 2015–2016, the LNAFC is decreased to 4342 from 5652 ha mainly because of massive deforestation in the district (Fig. 2.1). The maximum increase is noted in Ratua-I block (1275 ha) followed by English Bazar (802 ha). The maximum negative growth is reported in Habibpur block (789 ha). The land size category wise information represents an increase of 293 and 60 ha under marginal and small land size class categories correspondingly. The medium and semi-medium land size classes report corresponding decline of 1150 and 492 ha from 1995–1996 to 2015–2016 (Fig. 2.3).

2.6 Cropping Pattern

The cropping pattern simply refers as proportion of area under different crops in a region at a point of time. The proportion area includes net sown or gross cropped area depending upon types of cropping practice in a region. In a region, if only one crop is cultivated in a year then it is net sown area, and if more than one crop grows on same land then gross cropped area is taken for cropping pattern measure.

However, the meaning and concept of cropping pattern differs from region to region depending upon researcher's view or objective of the study. After looking into Indian farming systems, Johl and Kahlon (1963) feel that the cropping pattern of an area is the outcome of trial and adjustments with respect of farm enterprises and practices. There are innumerable independent variables which are not fixed but determine cropping pattern; therefore, cropping pattern of any region keeps on changing according to change in influencing factors. On this line, Mandal and Ghosh (1963) have considered cropping pattern in a wider perspective. They have combined cropping pattern with diversification and specialization in agriculture. They also indicate the risk and loss associated with different forces of cropping pattern. A study on "Analysis of Cropping Pattern of Farm Families in Surat District" by Desai (1977) has considered cropping pattern as the proportion of area under individual crop to the net cultivable land not gross cropped area. The net cultivatable areas are those available for cultivation in total geographic area of a region. As Venkataramanan and Prahladachar (1980) have explained, the cropping pattern is not simply the relative allocation of area under individual crops but relative change in area over the years. If the relative change over the year has not happened, then particular cropping system is named unchanged cropping pattern, and if relative area has changed over the years, then it is termed as elastic cropping pattern.

The type of cropping pattern of a region is determined by the number of crops which grows over a period of time. At a time, if the net sown area produces only one crop, then it is called monoculture or mono cropping pattern. And if two or more than two crops are grown simultaneously, then it is known as mixed cropping. Sometimes, on the same field more than one crop is cultivated along rows where one row of main crop, three or four row other crops are grown at a time. This system of cropping is known as intercropping. To maintain soil fertility of a unit land, different crops are grown in planned succession. It is known as crop rotation. The crop rotation may be a 1-year rotation or 2 years or more than that.

The cropping pattern of any region has never been constant. It is the result of interaction between the natural and human environment over space. The following factors significantly affect cropping pattern of a region:

1. The physical aspects of a place such as soil, rainfall, temperature, natural calamities, and unexpected weather aberration affect the cropping pattern significantly. In the climate change era, the unexpected weather phenomena, early or delay in monsoon rainfall, cyclone, hailstorms, and weather aberration are vital decision-making aspects to Indian farmers in choosing a particular cropping pattern. In most situations, the physical environment actually reduces the choice of farmer to grow different crops altogether (Morgan and Munton 1971). The recent change in India's cropping pattern from food to non-food crops suggests the burden of natural environment on food crop cultivation (ICAR 2018).

2. The economic factor plays a critical role in the expected cropping pattern of a region. The internal and external demands of certain crops, income opportunities, inflation, and price policy of government are crucial dimensions to a particular cropping pattern and change therein (Zandstra 1981).

3. The social aspects such as literacy, health, land tenancy and tenure, farm size, and migration also play key roles in cropping pattern determination at household level.
4. The government policy also affects the cropping pattern of a region. The crop insurance against loss and damage of crops encourages farmers toward profitable cropping patterns (Wu and Brorsen 1995).
5. Sometimes a single factor is enough to change the cropping pattern entirely of a region. The social, technological, and demographic aspects have been changing the cropping pattern of India (Singh 2012). In most of the time, the cropping pattern of a region is closely influenced by historical, geo-climatic, political, socio-cultural, and economic factors (Hussain 1996a).

2.7   Methods to Measure Changing Cropping Pattern

To measure the change in cropping pattern, there are two different methods that we have:

1. The change in areal share of a crop between two points of time – percentage method
2. The amount of change in cropping pattern in a region – ranking method

To measure the changing cropping pattern of a particular region between two points of time, the first method is useful. The second method is useful if regional comparison is to be drawn. The second method is applied to measure total amount of change in cropping pattern; therefore this method is useful to calculate positive or negative change in proportion of area of a region within a region. In the present study, the first method is used on gross area share by different crop categories such as cereals, pulses, oilseeds, and others from 1995–1996 to 2015–2016 by land size group at block level. The second method is employed on individual crops such as paddy, wheat, potato, and others from 1995–1996 to 2015–2016 across land size categories at block level.

2.8   Changing Cropping Pattern in Malda District: Percentage Method

Based on season, crops of India are divided into three types: Kharif crop (sown in June–July and harvested in September–October), Rabi crop (sown in October–November and harvested in April–May), and Zaid crop (March to June between Rabi and Kharif seasons). In India, Agriculture Census (2015–2016) records as many as 191 types of crop under 12 major crop categories that are grown in 3 different cropping seasons. The cereal crops remain on top in terms of both area and production. During the same Census, cultivation of 91 crops has been listed under 11 crop categories from the state of West Bengal. In Malda District, a total of 87 crops have been listed under major 9 crop categories. Because of negligible share of floriculture, fodder and green manure and other crop categories in net sown area, the analyses on mentioned crops have been omitted in the present discussion. A list of crop category has been attached in the Annexure section. The spatial and temporal change of cropping pattern under major crop categories (see Annexure 2.1) have been discussed in the following section.

2.9 Changing Cropping Under Gross Cropped Area (GCA)

Gross cropped area (GCA) is "the total area sown once as well as more than once in a particular year. When the crop is sown on a piece of land for twice, the area is counted twice in GCA. This total area is also known as total cropped area or total area sown" (Agriculture Census).

The overall GCA in district increased from 1995–1996 to 2015–2016. The GCA of a region is increased when institutional, technological, and infrastructural innovation has improved. The maximum increase is reported under the marginal land size category across different blocks of the district. Figure 2.4 displays land size class wise change in GCA in which only marginal land class gains positively in two decades. The positive change in marginal land size class indicates that the area and innovation in the same land size class happened positively throughout 1995–1996 to 2015–2016. In the marginal land size category, the maximum and minimum change in GCA is reported from Kaliachak-III (27.26%) and Manikchak (2.47%) blocks in the district. The positive change is also noted in Kaliachak-I, Kaliachak-II, Manikchak, Ratua-I, Chanchal-II, and Harishchandrapur-I blocks under small land size groups due to technological and infrastructural development in agriculture. The other land size groups have reported negative growth in GCA. The maximum negative change is found in semi-medium class in Ratua-I (17.33%). In medium and large land groups, Ratua-I (16.73%) and Kaliachak-III (0.29%) locate negative maximum change.

2.9.1 Changing GCA Under Cereal Crop

The cereals are also known as staple crops as they are the main diet sources to billions of world population. Three cereals, namely, maize, rice, and wheat, have shared around 90% of world's total cereal production. It is the source of main essential calories, carbohydrates, and nutrition, and therefore it is the basic human consumption source. Among six major cereal crops in India, rice and wheat have a lion share in the total GCA. In Malda District, the share of GCA under cereal crops was

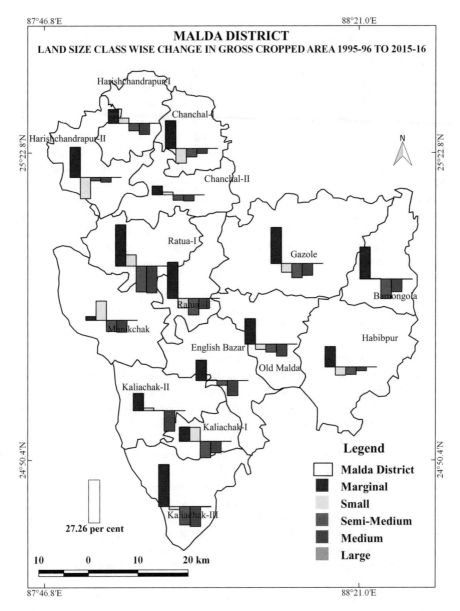

Fig. 2.4 Land size class wise change in gross cropped area in Malda District 1995–1996 to 2015–2016. (Source: See Annexure 2b)

72.31% which reduced to 63.01% in 2015–2016. Although the GCA under cereal crop has decreased but the number of crops in the same category increases to eight from six in two decades. The second half of the twentieth century has recorded impressive productivity of rice, wheat, maize, and other cereals. The bumper pro-

duction of cereals is reduced because most of the country has achieved self-sufficiency in food crop production. And therefore, the farmer brings extra area under
high-value crops.

Figure 2.5 reveals that the GCA under six land size groups reported negative
growth from 1995–1996 to 2015–2016. In the district, only Gazole block locates
increase in GCA under different land size classes. Except semi-medium and medium
land group, the other land size type reports positive growth in Kaliachak-II and Old
Malda blocks. The maximum negative change in GCA is stated in marginal
(25.45%), small (25.82%), all land size (23.90%), semi-medium (29.54%), medium
(31.28%), and large (19.45%) categories in English Bazar, Ratua-II, Chanchal-I,
and Kaliachak-III, respectively. This negative change is mainly because of risk from
natural calamities, price instability, and decreasing productivity of crops. The maximum positive change in GCA under different land size has been reported in
Kaliachak-II (20.14%) under small land group, in Kaliachak-II (7.69%) in marginal
land class, in Old Malda (4.90%) under semi-medium, in Ratua-I (3.44%) under
medium, and in Old Malda (4.49%) in all land size classes in the district.

2.9.2 Changing GCA Under Pulse Crop

The pulse crop is an important source of protein and fiber-rich food crop grown in
almost three different cropping seasons in district. Almost 80% of pulses are cultivated in Rabi cropping season in Malda District. Apart from medicinal use of pulses,
it is an important source of protein to almost half of India's population. Because of
growing demand of pulses in national and domestic markets, the GCA has increased
in India in the last two decades. However, national demand of pulses hardly falls
down, but here in Malda District, the GCA under same crops has decreased from
5.16% in 1995–1996 to 3.79% in 2015–2016. Except large land size groups, the
marginal, small, semi-medium, medium, and all land size categories have reported
negative change among which the maximum decrease in GCA is located in marginal land size group. The recurrent crop diseases, attack of insects, and low productivity have been important causes in decreasing GCA under pulse crop area in the
district. Although the GCA has decreased, the number of pulse crops rose from four
to seven in two decades. The diversification gets pushed from developed infrastructure in district.

Figure 2.6 shows that there are four blocks, namely, Chanchal-I (11.92%),
Harishchandrapur-II (11.67%), Chanchal-II (4.79%), and Bamangola (0.37%),
reporting positive growth in GCA under pulses in the district. In the same land size
class, Kaliachak-III (12.26%) and Manikchak (11.22%) blocks note negative
changes. The maximum positive change in GCA under small (14.72%), semi-
medium (9.69%), medium (0.96%), large (0.67%), and all land classes (11.21%) is
noted in Harishchandrapur-II, Chanchal-I, Old Malda, Kaliachak-III, and
Chanchal-I, respectively. The maximum negative change is found in Ratua-I

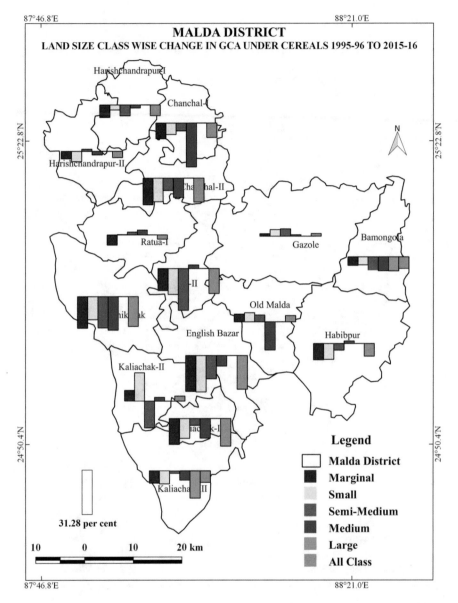

Fig. 2.5 Land size class wise change in GCA under cereals in Malda District 1995–1996 to 2015–2016. (Source: See Annexure 2c)

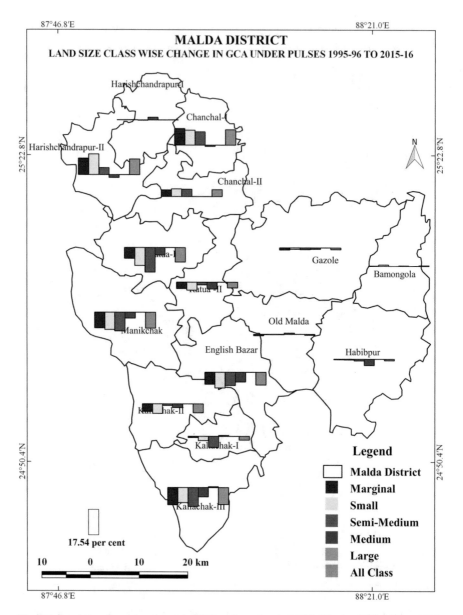

Fig. 2.6 Land size class wise change in GCA under pulses in Malda District 1995–1996 to 2015–2016. (Source: See Annexure 2d)

(12.83%), Ratua-I (17.54%), Kaliachak-III (6.67%), and Kaliachak-III (12.02%) under small, semi-medium, medium, and all classes in district.

2.9.3 Changing GCA Under Spice Crop

The spice crops have nutritional and medicinal benefits. That is why it is used as food crops and drug narcotics. It can be preserved and stored for a long time. The daily use of spices in different dishes of food has made it valuable. From 1995–1996 to 2015–2016, the number and GCA of spice crops have enlarged. In terms of share of GCA, chili, garlic, coriander, turmeric, and betel nut occupy highest to lowest in the ranking ladder. The number of spice crops rose to 15 from just 4 suggesting a booming diversification in the district. The increasing price hike of spice crops due to growing demand in national and domestic markets has played a key role in household decision-making in favor of these crops.

In 2015–2016, the marginal and all land size class reported positive gain in area in the district. In the marginal group, English Bazar block (1.64%) locates the maximum increase, while Gazole block (0.18%) notes the maximum negative growth. The positive growth is also found in small (0.39%), semi-medium (0.44%), and all land size groups (1.01%) in Kaliachak-III, Manikchak, and English Bazar block correspondingly. The maximum negative change depicts Chanchal-I (0.53%), Habibpur (0.70%), and Chanchal-I (0.11%) respectively in the district (Fig. 2.7).

2.9.4 Changing GCA Under Oilseed Crop

The oilseed crop is an important cash crop growing in almost every block under different land size classes in the district. The number and GCA under oilseed crops have increased in the district from 1995–1996 to 2015–2016. In the recent Census, the number of crops reported in the district is nine from six in 1995–1996. The major oilseed crops are mustard and rapeseed, linseed, sunflower, till, groundnut, coconut, and other oilseeds. The GCA under rapeseed and mustard has remained maximum in both the Census. The oilseeds are durable crops and have varieties of uses from households to industries in different food and non-food items. And therefore, the demand for oilseeds is increasing.

In district, except large land types, the other land group notes positive change in GCA. The maximum increase in GCA is noted under marginal class (6.12%) and then small (5.31), semi-medium (5.41%), medium (6.47%), and all land size class (5.78%) in the district. In marginal category, the maximum positive and negative changes have been reported in Habibpur (15.86%) and Kaliachak-II (−4.16%) blocks, respectively (Fig. 2.8). With 17.06% hike in Habibpur block and −0.371% decrease in Kaliachak-II block, overall, the small land size type has gained area in

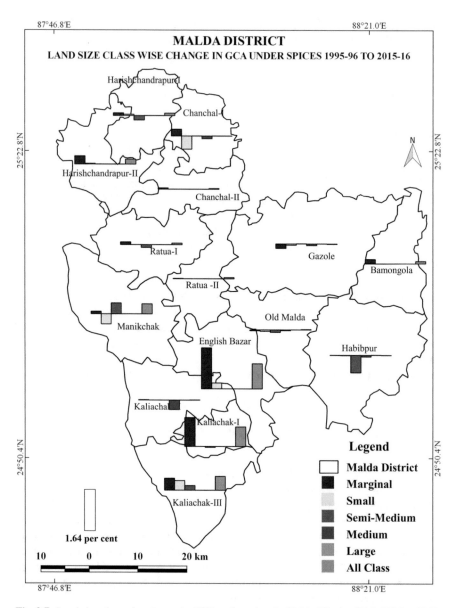

Fig. 2.7 Land size class wise change in GCA under spices in Malda District 1995–1996 to 2015–2016. (Source: See Annexure 2e)

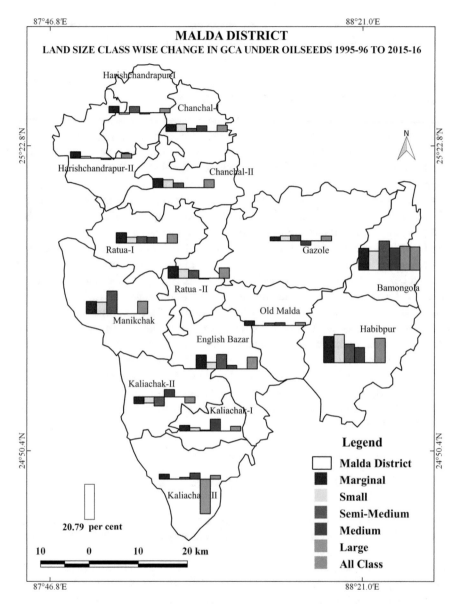

Fig. 2.8 Land size class wise change in GCA under oilseeds in Malda District 1995–1996 to 2015–2016. (Source: See Annexure 2f)

district. The maximum gains under semi-medium (17.47%), medium (13.28%), and large (14.43%) have been located in Bamangola block in the district.

2.9.5   Changing GCA Under Fiber Crop

The fiber crop is one of the types of non-food crops. Although the number of crops in 1995–1996 and 2015–2016 Census years remains the same, i.e., four, the GCA under it has increased across land size classes in the district. The jute, mesta, sunn hemp, and cotton are fiber crops grown in the district. The fiber crops are also known as industrial crops due to its vertical process in manufacturing industries or mills. Because of its durable nature and substitute product options, the demand of fiber crop in national and international market has been increasing.

In the district, the GCA under marginal land size class shows the maximum increase followed by all size class, small, semi-medium, and medium groups. The reason behind overall growth in GCA under fibers is it is risk-free and durable and can be stored for long. The maximum increase is recorded for Kaliachak -III in small (28.41%), semi- medium (22.44%), medium (15.24%), large (15.59%), and all land size classes (24.87%) among the blocks in the district. In the marginal category, the maximum increase is noted in Manikchak (24.28%), and the maximum negative growth is reported in Chanchal-II (9.37%) block (Fig. 2.9). The persistent highest growth is noticed in Kaliachak-III block under different land size classes. In Gazole block, the negative change in GCA remained throughout different land size classes. The maximum persistence in negative growth is recorded in Chanchal-II block in the district.

2.9.6   Changing GCA Under Vegetable Crop

The area under vegetable crops has decreased, but the number of crops in the same category has increased more than eightfold in the district. In the 1995–1996 Census, there were only three crops listed under vegetable crops. After two decades, the number of crops has reached 25 (see Annexure 2a). In both Censuses, the area under potato crop remains maximum followed by onion. After 2000, the number of vegetable crops has increased many folds, but the specialized crop has lost the dominance in areal share because of high market and diseases risk associated with it. The increasing diversification in vegetable crops is associated with growing demand for different crops in the local market. No doubt, vegetable cultivation is highly risk-prone crop in terms of its perishable nature, market demand, price instability, vertical linkages (with industries), natural calamities, and external factors altogether responsible in decreasing of GCA under vegetables.

In the district, the GCA under vegetables has reported negative change across different land size groups. In marginal (10.64%), small (9.09%), and semi-medium

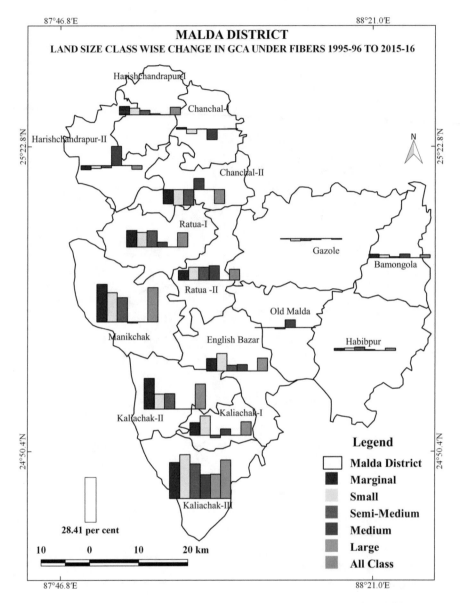

Fig. 2.9 Land size class wise change in GCA under fibers in Malda District 1995–1996 to 2015–2016. (Source: See Annexure 2g)

(8.64%) land size class, Bamangola block has reported the maximum negative growth in GCA. On the same line, English Bazar in medium (8.80%), Kaliachak-III (14.25%) in large, and Bamangola (8.80%) block have reported negative growth (Fig. 2.10). Among blocks, there are only two blocks that report positive gain in GCA. The maximum percentage of gain finds in Old Malda (11.31%) under medium

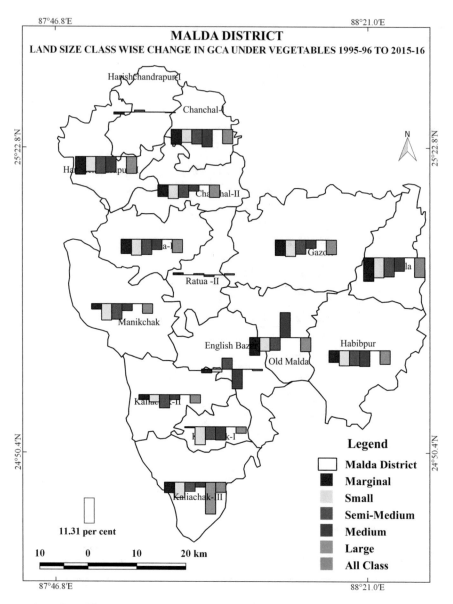

Fig. 2.10 Land size class wise change in GCA under vegetables in Malda District 1995–1996 to 2015–2016. (Source: See Annexure 2h)

category followed by Harishchandrapur-I (0.91%) block. The decreasing GCA under vegetable crops is unpredictable and has not remained the same in the past. In 2005–2006 and 2000–2001 Census year, the GCA under vegetables had found positive growth beyond different associated risks.

2.9.7   Changing GCA Under Fruit Crop

The number of fruit crops has grown up from 4 to 13 to 2015–2016 from 1995–1996. The increased number of fruit crops suggests that the district is moving toward fruit diversification. The area under the same crop category is also increased along with diversification. In the 2015–2016 Census, there are banana, guava, jackfruit, litchi, mandarin orange, mango, other citrus, papaya, pineapple, sapota, guava, temperate fruits, and miscellaneous fruits listed under fruit crops. In terms of share in GCA under fruits, mango occupies maximum area followed by litchi, banana, guava, and citrus fruits in both Census years. The demand of famous Malda mango has been increasing in national and international markets, and because of this, the area under mango crop has increased many folds in two decades. The area under litchi is increasing because its demand in the national market has increased.

In the district, the GCA under fruits has reported maximum gain under marginal land size type although other land size groups locate positive change except large land size category which displays negative change (Fig. 2.11). In marginal (23.45%), small (23.21%), and all land size category (22.13%), Chanchal-II block reports the maximum positive change, while Kaliachak-II (33.30%) and Chanchal-I (43.41%) have noted maximum positive growth under semi-medium and medium land size classes. In large land size class, both Bamangola and Kaliachak-III have noted negative growth, and this attributes in total negative change in district.

2.9.8   Changing GCA Under Total Food Crops

The food crops are those crops which are grown to feed humans in the form of seeds, grain, nuts, beverage, spices, vegetables, herbs, fruits, and others. The Agriculture Census publishes data on cereals, pulses, vegetables, fruits, and others as food crops. The production of food crops are the characteristics of subsistence agriculture especially in developing countries. In developed agriculture, food grain cultivation also serves the commercial purpose. From 1995–1996 to 2015–2016, the GCA under food crops decreased across land size classes in the district. The maximum negative change in GCA is reported under marginal land size class in 14 blocks (except Chanchal-II). The small, semi-medium, medium, large, and all land group class have noted negative growth in almost every block except Chanchal-I and Chanchal-II, Harishchandrapur-II, Old Malda, Manikchak, and Gazole blocks under noted land size categories (Fig. 2.12). The natural calamities, rainfall vari-

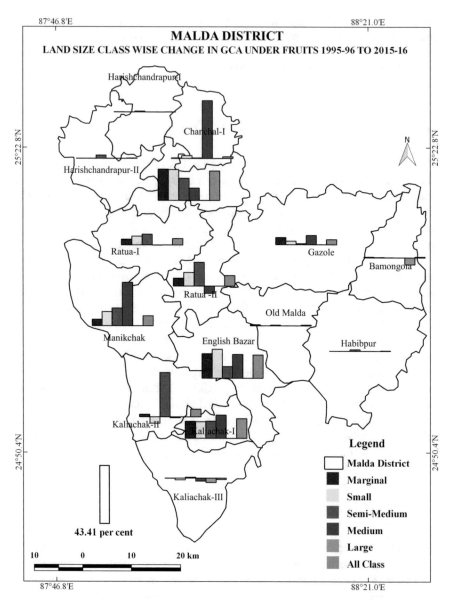

Fig. 2.11 Land size class wise change in GCA under fruits in Malda District 1995–1996 to 2015–2016. (Source: See Annexure 2i)

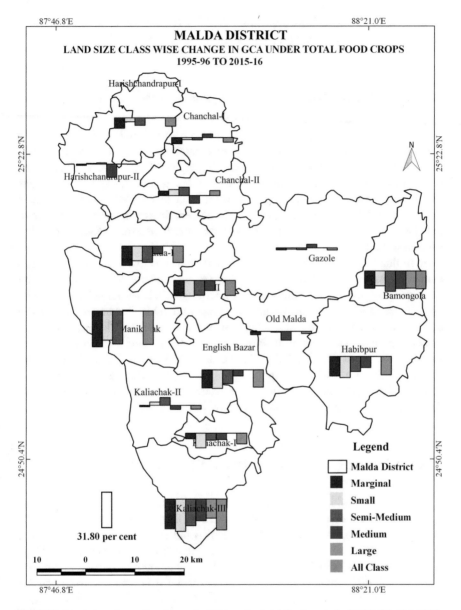

Fig. 2.12 Land size class wise change in GCA under total food crops in Malda District 1995–1996 to 2015–2016. (Source: See Annexure 2j)

ability, market infrastructure, price instability, and other external factors make agriculture risky especially food crops farming (Mahesh 1999).

2.9.9 Changing GCA Under Non-food Crops

In agriculture geography, non-food crops are also known as industrial crops because fiber crops, oilseeds, floriculture, sugar crops, bamboos, narcotic crops, green fodder, mulberry, rubber, and others are not directly used for human consumption; rather these are processed in industries for more valuable and alternate products. The recent change in cropping pattern from food to non-food crops is induced because it encourages new income avenues and employment. Agricultural innovation has increased crop diversification under non-food crops (Kalaiselvi 2012) because of its economic and social benefit.

In Malda District, the area under non-food crops has increased across different land size classes. The maximum addition of GCA is under marginal land class, but the maximum percentage gain is noted under large land size farmers. In marginal land size category, the maximum increase is reported in Manikchak block (31.80%), and only Chanchal-II block locates least and negative changes. The maximum positive change under small, semi-medium, medium, and all land size classes has been found in Kaliachak-III (28.52%), Manikchak (29.33%), Kaliachak-III (19.07%), and Manikchak (29.74%) correspondingly. The negative change in GCA shows in the Chanchal block under marginal, small, semi-medium, and all land size classes, while Chanchal-I reports maximum negative change under medium land group.

2.10 Changing Cropping Pattern: Ranking Method

The ranking method is used to calculate the amount of change in cropping pattern across land size groups in the district. In Malda District, there are 15 blocks. The ranking method helps explore total change in the cropping pattern of different blocks. The procedure of calculation of ranking method is mentioned in Table 2.1.

An attempt has been made to calculate the total amount of change in cropping pattern under different land size classes in the district. To do so, the first amount of change has been calculated for Bamangola block under marginal land size category from 1995–1996 to 2015–2016. Table 2.1 shows that ten crops have occupied 100% gross cropped area in 1995–1996, and then the rank has been assigned against the areal share of crop under total GCA in ranking column. In 2015–2016, the area of same ten crops has decreased to 99.67%. Against the share of individual crops in total GCA, the rank of individual crop has been updated. In 2015–2016 ranking

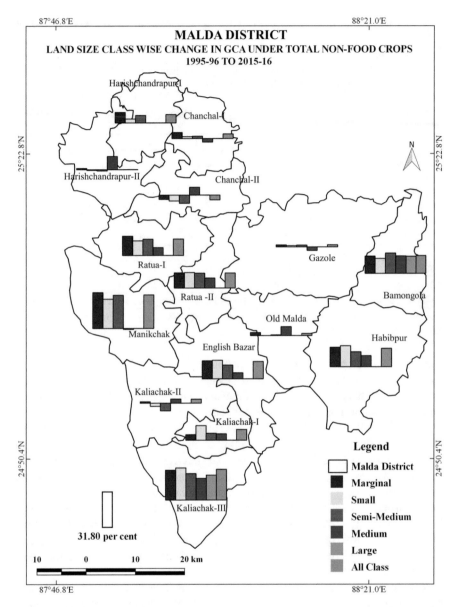

Fig. 2.13 Land size class wise change in GCA under total non-food crops in Malda District 1995–1996 to 2015–2016. (Source: See Annexure 2k)

Table 2.1 Total change in cropping pattern for Bamangola block in marginal category, 1995–1996 to 2015–2016

Bamangola block						
	Percentage share in total GCA of block		Ranking of the crop against area		Change in ranking	
Crops (1995–1996 as base year)	1995–1996	2015–2016	1995–1996	2015–2016	1995–1996 to 2015–2016	Total change (avoid negative sign)
Paddy	72.94	64.96	1	1	0	0
Mustard	10.42	25.37	2	2	0	0
Other vegetables	9.14	2.29	3	10	−7	7
Wheat	4.36	2.36	4	3	1	1
Potato	2.54	2.11	5	4	1	1
Linseeds	0.52	0.59	6	52	−46	46
Other fruits	0.04	0.50	7	41	−34	34
Tur	0.02	0.30	8	55	−47	47
Jute	0.02	0.11	9	5	4	4
Chili	0.00	0.09	10	9	1	1
Total area share	100.00	99.67	**Total change in cropping pattern**			**141**

Source: Agricultural Census, Government of India, 1995–1996 and 2015–2016 (based on author calculation)

column, since the number of crops is increased from 16 to 31 in mentioned block, therefore, the ranking order has also changed from 1995–1996 to 2015–2016. However, any change in proportional share of crops is considered as changing cropping pattern according to percentage method.

However, if the proportional share of crop is changed but the rank remains the same, then it shall not be taken as changing cropping pattern. Sometimes, the areal share of a crop remains the same in both years but the rank has changed. Then also the changing cropping pattern is experienced. In changing ranking column, if the value is zero meaning thereby there is no change in rank of that particular crop. In case of negative change, the rank of crop decreases, while positive change increases the rank positively. Here, the author is interested to know the total amount of change in cropping pattern regardless of positive and negative; therefore, by use of absolute modulus function, the negative sign is avoided in the last column of the mentioned table.

The total amount of change of cropping pattern is calculated among different land size types from 1995–1996 to 2015–2016. Table 2.1 shows that the amount of change under marginal land size class category is higher than the rest of land size categories. Even in small land size class category, the amount of change is much closer to marginal land size but less than semi-medium and higher than medium land size groups. The maximum amount of change is noted under the marginal land size category in Gazole (176) block, and the minimum change is reported in Manikchak (17) block under semi-medium land size type. A study by De and Chattopadhyay (2009) has concluded that if a family consists of four members, out

Table 2.2 Total change in cropping pattern under different land size class categories in Malda District (1995–1996 to 2015–2016)

Block	Marginal	Small	Semi-medium	Medium	Large	All class
Bamangola	141	109	67	81	79	130
Chanchal-I	156	115	126	90	N. A	150
Chanchal-II	169	136	112	84	N. A	169
English Bazar	130	97	73	85	N. A	128
Gazole	176	135	122	92	N. A	174
Habibpur	168	140	143	108	N. A	167
Harishchandrapur-I	134	94	93	88	N. A	131
Harishchandrapur-II	121	88	65	67	N. A	116
Kaliachak-I	142	167	80	75	N. A	133
Kaliachak-II	121	85	126	81	N. A	122
Kaliachak-III	81	80	30	42	50	79
Manikchak	82	41	17	69	N. A	80
Old Malda	98	48	45	58	N. A	96
Ratua-I	104	80	74	53	N. A	103
Ratua-II	166	141	93	117	N. A	159
District	60	50	46	111	49	59

Source: Computed by author
N. A – Not available

Table 2.3 ANOVA summary

ANOVA summary						
Source of variation	SS	df	MS	F	P-value	F crit
Between groups	40755.45	5	8151.09	8.225174	0.000004	2.34368
Within groups	70360.5	71	990.993			
Total	111115.9	76				

Source: Computed by author

of them one at least is a migrant (15–59 age group). The remittances from migrant workers have added in total family income as outsourcing. The family earning from outsourcing significantly impacts decision-making of expected cropping patterns as it would whether be low-value to high-value market crops. Therefore, marginal farmers are very frequently into changing their cropping pattern (Table 2.2).

The first hypothesis of the study is that there is no significant difference in the amount of change of cropping pattern across different land size categories.

Table 2.3 reveals that between groups (land size classes) and within groups (blocks) variations are not indifferent. The high F statistics (8.22) and F-critical (2.34) values are statistically significant as P-value less than 0.001 confirms that the amount of change in cropping pattern under land size classes is different.

2.11   An Insight from Field Survey

The primary survey shows that 84% of farmers of the district change their cropping pattern very often (mean less than 3 years). Out of total farmer in marginal land size class category, 86% of respondents address that they are changing their cropping pattern in less than 3 years crop-wise as well as season-wise. The small farmers have been changing cropping pattern because of profit criteria. About 82% of small farmers agree that they favor change in cropping pattern at least once in 3 years. The semi-medium farmers have changed cropping pattern due to consumption and profit. About 67% of farmers do change cropping pattern from low- to high-value crops like vegetables and horticultures. The rate of change in cropping pattern is comparatively low among medium land size farmer. Around 60 percent medium land size farmer changed cropping pattern because of certain risk factors. The amount of change in cropping pattern is higher under vegetable, pulses, and oilseeds under marginal and small landholding farmers. The medium and semi-medium land size farmers have mostly changed their crops from cereals to fibers and in some cases fruits and oilseeds (Fig. 2.14).

The reasons for changing cropping patterns have been presented in Fig. 2.15. In field survey questionnaires, the nature of option was closed ended, and therefore other than mentioned four options were merged with most related one. Figure 2.15 illustrates that the profit opportunity option plays a key role at households in decision-making to change cropping patterns. The semi-medium and small land size groups are more inclined toward profit option than other classes. Because of small land size, the marginal and small farmers prefer edible food crop cultivation

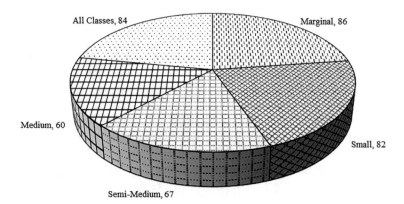

Fig. 2.14 Percentage of farmers who change cropping pattern very often in Malda District 2018. (Source: based on field survey 2018)

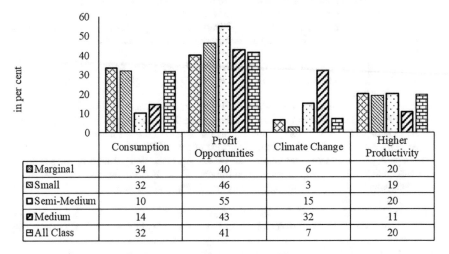

Malda District
Reasons of Changing Cropping Pattern 2018

	Consumption	Profit Opportunities	Climate Change	Higher Productivity
Marginal	34	40	6	20
Small	32	46	3	19
Semi-Medium	10	55	15	20
Medium	14	43	32	11
All Class	32	41	7	20

Fig. 2.15 Reasons for change in cropping pattern in Malda District 2018. (Source: based on field survey 2018)

than profit opportunities. The type of agricultural practice in the district is subsistence cum slow commercial, therefore the cultivation of edible food crop for consumption is another important factor in household decision-making. The semi-medium and medium land size classes have been changing cropping patterns comparatively lower than other land size groups. The higher productivity of crop is another decision-making option which favors change in cropping pattern more among marginal and semi-medium land groups than other groups. The variability in monsoon rainfall, drought, weather aberrations, and natural calamities attribute in climate change which affect medium and semi-medium land size classes more as compared to marginal and small land groups.

References

Beale, L. (1997). An inventory of Europe-wide land use and land cover studies. The User Needs for More Harmonized Land Use Information at the National and the EU Level.

Bourne, L. S. (1982). Urban spatial structure: An introductory essay on concepts and criteria. *Internal Structure of the City, 45*.

Briassoulis, H. (2019). *Analysis of land use change: Theoretical and modeling approaches* (No. 333.76 B849) (Vol. 3, pp. 13–16). Morgantown: Regional Research Institute, West Virginia University.

Christian, C. S. (1957). The concept of land units and land systems. *Proceedings Ninth Pacific Science Congress, 20*, 54–59.

Comprehensive District Agricultural Plan (C-DAP) (2015). *Malda–draft* report. Uttar Banga Krishi Viswavidyalaya. Pundibari, Coochbehar, West Bengal.

Desai, B. M. (1977). Analysis of cropping pattern of farm families, Surat District. *Indian Journal of Agricultural Economics, 32*(902-2018-1334), 78–91.

De, U. K., & Chattopadhyay, M. (2009). Crop diversification by poor peasants and role of infrastructure: Evidence from West Bengal. *Journal of Development and Agricultural Economics, 2*(10), 340–350.

Food and Agriculture Organization (FAO). (1950). *Planning for Sustainable Use of Land Resources. FAO Land and Water Bulletin 2.* Rome: Food and Agriculture Organization of the United Nations.

Food and Agriculture Organization of the United Nations. Land, Water Development Division, & Agriculture Organization of the United Nations. Interdepartmental Working Group on Land Use Planning. (1995). *Planning for sustainable use of land resources: towards a new approach* (No. 2). Rome: Food & Agriculture Organization of the UN (FAO).

Food and Agriculture Organization of the United Nations. Soil Resources, Management, Conservation Service, Agriculture Organization of the United Nations. Land, & Water Development Division. (1996). *Agro-ecological zoning: Guidelines* (No. 73). Rome: Food & Agriculture Organization.

Graetz, D. (1994). "Grasslands". In *Changes in Land Use and Land Cover: A Global Perspective*, Edited by: Meyer, W. B. and Turner, B. L. I. 125–147. Cambridge: Cambridge University Press

Hoover, E. M., & Giarratani, F. (1984). *An introduction to regional economics* (3rd ed.). New York: Alfred Knopf. Reprinted in Loveridge, S. (Ed.) (1999). *The web book of regional science* (pp. 131). Morgantown: West Virginia University, Regional Research Institute.

Hussain, M. (1996a). *Systematic agricultural geography.* New Delhi: Rawat Publications.

Hussain, M., (1996b). *Systematic agricultural geography,* Reprinted 2007, Rawat Publication, Jaipur/New Delhi, pp. 217, 218.

Indian Council of Agricultural Research (ICAR) (2018). *Annual Report.* Department of Agricultural Research and Education. Ministry of Agriculture and Farmers Welfare. Government of India

Johl, S. S., & Kahlon, A. S. (1963). Economics of cropping pattern (an analytical case study). *Indian Journal of Agricultural Economics, 18*(902-2016-67012), 132–142.

Jones, A., & Clark, J. (1997). Driving forces behind European land use change: An overview. In *The user needs for more harmonized land use information at the national and EU level.* (Report on the CLAUDE, 24–31).

Kaiser, E. J., Godschalk, D. R., & Chapin, F. S. (1995). *Urban land use planning* (Vol. 4). Urbana: University of Illinois Press. Ch. 4.

Kalaiselvi, V. (2012). Patterns of crop diversification in Indian scenario. *Annals of Biological Research, 3*(4), 1914–1918.

Kleckner, R. L. (1981). A National Program of land use and land cover mapping and data compilation. *Planning Future Land Uses, 42,* 7–13.

Mahesh, R. (1999). *Causes and consequences of change in cropping pattern: A location-specific study* (p. 56). Kerala Research Programme on Local Level Development, Centre for Development Studies.

Mandal, G. C., & Ghosh, S. K. (1963). Some aspects of the economics of cropping pattern a study of condition in the district of Monghyr, Bihar. *Indian Journal of Agricultural Economics, 18*(902-2016-67014), 74–83.

Memoria, C.B. (1984). *Agricultural Problems of India.* Kitab Mahal, Allahabad, p.93.

Meyer, W. B. (1995). Past and present land use and land cover in the USA. *Consequences, 1*(1), 25–33.

Meyer, W. B., & BL Turner, I. I. (Eds.). (1994). *Changes in land use and land cover: a global perspective* (Vol. 4). Cambridge University Press

Meyer, W. B., & Turner, B. L. (1996). Land-use/land-cover change: Challenges for geographers. *GeoJournal, 39*(3), 237–240.

Mohammad, N. (1992). *New Dimensions in Agricultural Geography: Dynamics of agricultural development* (Vol. 7). New Delhi: Concept Publishing Company.

Morgan, W. B., & Munton, R. J. C. (1971). *Agricultural geography*. London: Methuen.

Moser, S. C. (1996). A partial instructional module on global and regional land use/cover change: Assessing the data and searching for general relationships. *GeoJournal, 39*(3), 241–283.

Pierce, T. W., & Thie, J. (1981). Land inventories for land use planning in Canada. *Planning Future Land Uses, 42*, 57–71.

Singh, G. (2012). Factors influencing cropping pattern in Bulandshahr district-with special reference to the size of land holding. *International Journal of Scientific and Research Publications, 2*(5), 1–10.

Turner, B. L., & Meyer, W. B. (1994). Global land-use and land-cover change: an overview. *A Changes in land use and land cover: a global perspective, 4*(3).

Turner, B. L., Meyer, W. B., & Skole, D. L. (1994). Global land-use/land-cover change: Towards an integrated study. *Ambio Stockholm, 23*(1), 91–95.

Turner, B. L., II, Skole, D., Sanderson, S., Fischer, G., Fresco, L., & Leemans, R. (1995a). *Land-use and land-cover change: Science/research plan* (IGBP Report No. 35, HDP Report No. 7). Stockholm/Geneva: IGBP and HDP. Ch. 22.

Turner, B. L., II, Skole, D., Sanderson, S., Fischer, G., Fresco, L., & Leemans, R. (1995b). *Land-use and land-cover change; science/research plan* (IGBP Report No. 35, HDP Report No. 7). Stockholm/Geneva: IGBP and HDP. Ch. 20.

Venkataramanan, L. S., & Prahladachar, M. (1980). Growth rates and cropping pattern changes in agriculture in six states: 1950 to 1975. *Indian Journal of Agricultural Economics, 35*(902-2018-1665), 71–84.

Wolman, M. G. (1987). *Criteria for land use. Resources and world development* (pp. 643–657). New York: Wiley.

Wu, J., & Brorsen, B. W. (1995). The impact of government programs and land characteristics on cropping patterns. *Canadian Journal of Agricultural Economics/Revue canadienned'agroeconomie, 43*(1), 87–104.

Young, A. (1975). Rural land evaluation. *Evaluating the Human Environment, 5*–33.

Zandstra, H. G. (1981). *A methodology for on-farm cropping systems research*. Los Baños: International Rice Research Institute.

Chapter 3
Agriculture Development: Inputs-Outputs Dimension

In the previous chapter, we have observed that the land use and cropping pattern dynamics differ under various land size groups in the district. When the basic unit of agriculture (i.e., land) changes at spatiotemporal context, then other inputs such as infrastructure, technologies, government policy, and others do contribute to amplify the change positively of the whole agricultural system. Against this background, if the agricultural situation of a region is to be measured, then both aspects like inputs and outputs have to be taken into consideration. In the present chapter, input (both physical and non-physical determinants) and output dimensions of agriculture have been examined by using time series and decadal data. Next to the theoretical aspects of agricultural development, the trends and pattern of agricultural inputs (physical and non-physical) and outputs (productivity and crop diversification) have been examined to know which input plays a major role in agricultural output. In the following sequence, the non-physical determinants and their role have been measured. This sequence further extends to know the nature of association between crop diversification and non-physical inputs of agriculture. In the last section, the overall development of agriculture concerning inputs and outputs has been discussed at spatial context across the land class categories in two points of time at block level.

3.1 Agricultural Development: Concept and Importance

The word agriculture includes art and science of cultivation of land for crops and livestock. The land cultivation and livestock rearing already have independent academic discourse as agronomy and science of animal husbandry. Agriculture as a discipline relates with biological science, economics, sociology, public policy, statistics, geography, and other disciplines. On the other hand, development as a word means a process of progressive change, growth, and advancement of

H. Rahaman, *Diversified Cropping Pattern and Agricultural Development*,
https://doi.org/10.1007/978-3-030-55728-7_3

something. Though the word 'development' is basically a concept of economics but other disciplines are too linked with it. Therefore, both agriculture and development as academic discourse are very multidimensional and, hence, complex issues together to measure with universal acceptance. At this point agricultural development must be understood in an integrated fashion because as an activity it has direct link with lives and livelihood, economic growth, and development.

Importance of agricultural development for social causes bears much importance to third world countries. A study done by Jathar and Beri (1949) concluded that agricultural development is necessary to improve the social life of rural masses, to ensure food supply, and to promote agro-based industries. The continuous decline of landholding limits farm mechanization which makes it difficult especially for cereals and commercial crops. To tackle social problems, agricultural extension becomes necessary. Johnston and Mellor (1961) have examined the interrelationship between agriculture and industry to comprehend the nature of socioeconomic growth. The study found that an increasing agricultural output meets the demand of expanding food, increases foreign exchange, and helps in the expansion of industries. On theoretical consideration and historical experience, study suggests the financial investment to spell out stagnant agricultural expansion. Through agricultural development, an attempt has been made "to develop social capital, neighborly relations and mitigate social constraints created by non-farmers at rural-urban interface" (Pretty and Ward 2001). The present agricultural progress counts more in economic sense than social, cultural, and other aspects. Therefore, in economic growth and development, what gets paid from agricultural development has become an important policy concern.

3.2   Growth-Development Nexus: Agriculture

With respect to agricultural development, the serious unsolved question of whether development leads growth or growth pushes development remains critical before policy analysts and academia. However, India's agricultural policies have been much inclined toward development-oriented growth evinced from different 5-year plans. From the first 5-year plan to present sectoral plan, agricultural expenditure witnessed development purpose investment due to engagement of huge size of rural population rather than its 16.7% (India 2019) share in national GDP. India's GDP drains from other sectors more than agriculture; therefore, growth of the national economy is much dependent on non-agricultural activities. To push up rural and national economy, the institutional and infrastructural development becomes necessary especially from gained earnings of non-agricultural sectors. As a responsibility of the welfare state, development for unskilled rural peasants is a much-needed step toward social justice. Therefore, measurement of growth and development linkage must be reviewed through sound methodologies.

Methodologically, measuring agricultural development enjoys two separate approaches, namely, qualitative and quantitative. Qualitative dimension explains in Chap. 6 based on field survey. Here, taking notes from popular and accepted viewpoints, quantitative dimensions have been analyzed through the best accessible and available data.

As we see in the introductory part of this book, the assessment of agricultural development is based on inputs and outputs dimensions. Here, those different dimensions discuss through data in the following section.

3.3   Inputs-Outputs Determinants of Agriculture

Both physical and non-physical factors influence agriculture. Important physical factors are soil nutrients (include macro and micro), temperature, rainfall, and humidity, while non-physical inputs comprise land tenancy, irrigation, seeds, mechanization, fertilizers, credit facilities, and socioeconomic aspects (literacy, age, family size, workforce, and others). In the present study, two agricultural outputs (productivity and crop diversification) have been taken for discussion. To get a holistic picture on trends and patterns of inputs-outputs and their relation, this present section is a way forward. Here relationship has been assessed with inputs (physical as well as non-physical) and output determinants of agriculture through different statistical tools and techniques.

3.3.1   Physical Determinants of Agriculture

A set of 40 variables under 5 indicators have been selected as physical determinants of agriculture in present analyses. The selection of variable sources from extensive literature survey. Because of the limitation of disaggregate data on physical indicators at the block level for 1995–1996 and 2015–2016, the present section of the study emphasizes on time series data to assess trends, pattern, and relation of agricultural inputs-outputs. The importance of physical indicators in agriculture can be understood in the following ways:

1. Rich soil nutrients (macro and micro) always help groom crops productivity hence agriculture output. Decreasing soil nutrients is a kind of disease on the health of the crop (Barber 1995).
2. Increasing temperature is a way forward of sustainability of plants which are at the verge of extinction (Theurillat and Guisan 2001). Optimum temperature in different cropping seasons is necessary for better agricultural outputs.
3. Amount of rainfall does not help in sources of irrigation exclusively, but natural rainfall is essential for healthy growth of the plants (Bhuiyan and Kogan 2010).
4. Seed germination in humid countries is often problematic due to the amount of persistent humidity (Körner and Challa 2003).

Table 3.1 List of physical indicators and variables of agricultural development

Ind.	Var.	Description of variable	Ind.	Var.	Description of variable
Macro soil nutrients	X1	Number of acidic soil sample (pH scale)	**Micro soil nutrients**	X20	Sodium (S) sufficient sample
	X2	Number of neutral soil sample (pH scale)		X21	Zinc (Zn) sufficient sample
	X3	Number of alkaline soil sample (pH scale)		X22	Iron (Fe) sufficient sample
	X4	Electrical conductivity <1 sample		X23	Copper (cu) sufficient sample
	X5	Electrical conductivity 1–2 sample		X24	Manganese (Mn) sufficient sample
	X6	Electrical conductivity 2–3 sample		X25	Boron(B) sufficient sample
	X7	Electrical conductivity >3 sample	**Temperature**	X26	Average maximum temperature in Kharif season
	X8	Percentage of low organic carbon (in sample number)		X27	Average maximum temperature in Rabi season
	X9	Percentage of medium organic carbon (in sample number)		X28	Average maximum temperature in Zaid season
	X10	Percentage of high organic carbon (in sample number)		X29	Average minimum temperature in Kharif season
	X11	Low nitrogen (N) content sample		X30	Average minimum temperature in Rabi season
	X12	Medium nitrogen(N) content sample		X31	Average minimum temperature in Zaid season
	X13	High nitrogen (N) content sample	**Rainfall**	X32	Average rainfall during Kharif season (in mm)
	X14	Low phosphorus (P) content sample		X33	Average rainfall during Rabi season (in mm)
	X15	Medium phosphorus (P) content sample		X34	Average rainfall during Zaid season (in mm)
	X16	High phosphorus (P) content sample	**Humidity**	X35	Maximum relative humidity during Kharif season
	X17	Low potash (K) content sample		X36	Maximum relative humidity during Rabi season
	X18	Medium potash (K) content sample		X37	Maximum relative humidity during Zaid season
	X19	High potash (K) content sample		X38	Minimum relative humidity during Kharif season
				X39	Minimum relative humidity during Rabi season
				X40	Minimum relative humidity during Zaid season

Note: Original data on the above variables and indicators attached in annexures 3a–3e

Data on abovementioned variables has been standardized, and composite index of indicators has been calculated through the following technique:

(i) For positive indicator/variables of development

$$\frac{\left(\text{Actual value} - \text{Minimum Value}\right)}{\left(\text{Maximum value} - \text{Minimum Value}\right)}$$

(ii) For negative indicator/variable of development

$$\frac{\left(\text{Maximum value} - \text{Actual Value}\right)}{\left(\text{Maximum value} - \text{Minimum Value}\right)}$$

If data is in percentage form, then the minimum value should be zero (0) and the maximum should be hundred (100). In case, data is in absolute number, for that minimum and maximum value should be from table value. These techniques are also used in Chap. 4.The composite standardized score (CSS) of physical indicators is depicted in Fig. 3.1. It shows that the line graphs of four indicators (macronutrients, micronutrients, rainfall, and humidity) display a decreasing trend, while the temperature index is increasing. The macro soils nutrient index which includes soil pH, organic carbon, nitrogen (N), electrical conductivity, phosphorus (P), and potassium (K) is dropping their quality and quantity both (see Annexure 3a–3e).

The pH stands for "potential of Hydrogen." It is simply a measure of the binding relationship between hydrogen and hydroxyl ions. The pH value measures on a scale which ranges from 0 to 14. If pH value in scale is below, 7 then it is acidic, and if the

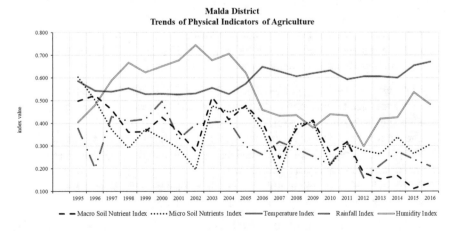

Fig. 3.1 Trends of physical indicators of agriculture in Malda District. (Source: based on Annexure 3a–3e)

same is more than 7, then it is alkaline. If the pH value of soil is 7, it is neither acidic nor alkaline; it is good for cultivation. The potential hydrogen in soil necessarily influences the availability of other essential nutrients of soil. Most crops grow within a suitable range of pH that is slightly acidic to slightly alkaline or 6 to 7.5 pH scale. In 1995–1996, the pH scale of soil showed 46.85% acidic, 27.28% neutral, and 25.90% alkaline that changed into 41.13, 35.73, and 23.16% in 2015–2016, respectively. It seems neutral soil percentage went up because of the increase in sample size. Otherwise, there would've been a variation on overall trends.

The electrical conductivity (EC) is another dimension of soil quality. It represents overall quality of soil especially physical and chemical characteristics which is determined by soil color, texture, porosity, density, consistency, structure, temperature, colloids, cation capacity, and others (Batjes 1997). To cultivate crops, EC helps in many ways and also determines the function of other organic and inorganic materials present in soil. The EC of soil depends also upon soil particles and moisture content in it. The sandy soil has low EC as this contains low moisture than silt and clay soil. The EC of clay is all time high. The ability to conduct electric current is expressed in a unit of desiSiemens per meter (dS/m). On ds/m scale, if value displays less than 1, then it is normal soil, while $1–2$ dSm^{-1} is critical for germination, $2–3$ dSm^{-1} is considered critical for salt-sensitive crops and above 3 indicates very injurious to most of the crops (Robbins et al. 1991). Annexure 3a shows that more than 97% soil in the district is under less than dSm^{-1} which means in normal category.

The soil has organic matter composed of carbon, hydrogen, and oxygen and has small amounts of other elements. Organic matter of soil helps develop the ecologically sound agricultural practice of a region. The presence of organic matter in good soil remains high, and it starts to decrease as human intervention increases in the form of addition of inorganic inputs in soil. Annexure 3a displays around 49.34% soil of the district having low organic carbon, while 39.70% medium and 11.00% low in 1995–1996. In 2015–2016, organic carbon is reduced to 52.39, 38.81, and 8.80% under low, medium, and high category, respectively.

In the overall life process, the balanced nitrogen (N), phosphorus (P), and potassium (K) ration are essential. The NPK ratio is the basic life cycle requirement of plants. These are helpful in many ways:

1. Nitrogen is an essential component of chlorophyll. It affects the nature of trees, its leaves, seeds, and fruits. In overall growth and development of a plant, adequate nitrogen is essential.
2. The nature of plant roots is directly influenced by the phosphorus content of soil. The plant oil, sugar, and starch which are generated in the plant body are mainly encouraged from adequate phosphorus.
3. Potassium helps in plant growth, buildups protein stores, and also aids in preventing diseases through increasing plan immunity.

It is revealed from Annexure 3b that about 69.80% of soil suffered from low nitrogen intact, 29.76% from medium, and just only 0.40% soil had high nitrogen content in 1995–1996 which went down further in 2015–2016 as 76.82, 22.86, and 0.36%

under low, medium, and high nitrogen accordingly. In the same period, phosphorus content of soil is also found decreasing trends with temporal variations in it. As it were 54.87%, 24.76%, and 20.40% in 1995–1996 and reduced to 40.97%, 24.01%, and 10.18% in 2012–2013 under low, medium, and high category, respectively. Potassium content of soil stood as 30.62%, 43.35%, and 26.06% in 1995–1996 which is changed into 39.22, 44.91, and 15.86 in 2015–2016 under low, medium, and high group.

Decreasing macro soils nutrient is a serious concern for productivity although it could be raised through high doses of fertilizers and other agriculture inputs; again, it causes multiple threats. Under macro soil nutrient index, organic carbon is also found in a decreasing trend (see Annexure 3b). Though the district has a normal electrical conductivity of soil, recent soil health card data shows that the percentage of alkaline soil is increasing due to overdoses of fertilizers.

The CSS of micro soil nutrients includes sodium (S), zinc (Z), manganese (Mn), iron (Fe), copper (Cu), and boron (B). In 1995, the micro soil index stood 0.497 which decreased to 0.137 in 2016. In the growth and development of plants, the micronutrients like sodium, zinc, iron, copper, and manganese play key roles.

1. In plants, zinc is crucial for the sake of protein and enzyme. Deficiency of zinc reduces the quality and yield of plants.
2. Manganese supports reasonable plant growth at different stages of life cycle. In the daily biological life system, manganese helps in various ways like photosynthesis, respiration, pollen germination, nitrogen assimilation, development of root cells, and making it disease resistant.
3. Iron and copper also are essential soil nutrients which support plant growth at a reasonable rate.
4. Any deficiency of boron is adversely affecting the physiological function including growth and development of plants. The boron directly helps in structural and functional activity through creation and maintenance of cell wall, cell membrane, cell division, and other internal functions of plants.

Micro soil nutrients index reports decreasing trends from 1995 to 2016 in the district (Fig. 3.1). In 1995 the CSS index stood as 0.497 which decreased to 0.427 (2000), with slight increase in 2005 (0.478) and again decreased to 0.137 in 2016.

Temperature index observes increasing variability over 20 years in the district. Increasing temperature indicates, maximum mean temperature during Kharif, Rabi, and Zaid seasons is going up. In 1995 the temperature index reported 0.546 that rose into 0.670 in 2016. Though there is variation in temperature graph (Fig. 3.1), the overall situation suggests an increasing trend. As we know plants gain growth during night time and if the temperature index continues to increase for a long time, then it harms the plant's growth and thus productivity of the crop. The increasing temperature leads plants' health to be vulnerable.

Rainfall graph during different cropping seasons reports a decreasing trend in the district (Fig. 3.1). Average annual rainfall has been decreasing (as we also see in Chap. 1) over the past 20 years. The district has already experienced a declining groundwater table, and if it is continuing, in the long run, it not only reduced irrigation sources; rather the production of crops would be difficult. Figure 3.1 depicts that the rainfall index (from Kharif, Rabi, and Zaid) went down from 0.378 in 1995 to 0.211 in 2016. More than 65% of crop cultivation is based on monsoon rainfall; therefore, in such cases, the decreasing rainfall is an unyielding concern to agricultural output.

Figure 3.1 shows reasonable variability in the humidity index. It helps in seed germination and plant growth. Although the amount of rainfall has decreased, the amount of relative humidity in the air remains almost the same through different years. In 1995, humidity index (from Kharif, Rabi, and Zaid season) recorded as 0.405, which reached to 0.483 in 2016. Increasing relative humidity although doesn't guarantee the occurrence of rainfall. As of now, humidity has been normal.

3.3.2   Non-physical Determinates of Agricultural Development

In this study, a total of 32 input variables and 2 output variables have been carefully chosen based on extensive literature survey. A study done by Joshi and Dube (1979) considered combined rank score to measure agricultural development. A set of seven indicators, namely, irrigated area, cropping intensity, productivity, commercialization degree, number of tractors, cattle density, and net sown area are taken into consideration. An analysis on development and productivity of Indian agriculture by Tewari and Singh (1985) has selected eight indicators for the construction of composite index. These are:

1. Value of produce per hectare NSA
2. Fertilizer consumption in per hectare cropped area
3. Irrigation coverage
4. Irrigation intensity
5. Number of agricultural workers in per hectare NSA
6. Cropping intensity
7. Percentage of commercial crops to GCA
8. Consumption of power per hectare of NSA

Singh (1987) has done a study on "an analysis of correlation matrix in agricultural development: a case study of Mirzapur district." He has selected seven indicators:

1. Cultivable area per agricultural worker
2. Net sown area per agricultural worker
3. Area sown more than once as percentage to NSA
4. Net irrigated area to net sown area
5. Wage rates of agricultural worker
6. HYV area to NSA
7. Fertilizer consumption per hectare of cultivable land

A study conducted by Bhadrapur and Naregal (1992) includes 18 indicators from land tenancy, irrigation, fertilizer, farm technologies, credit facilities, price, cropping intensity, productivity, nature of workers, and land use aspects. The study concluded that environmental, social, physical, and cultural variations lead to regional disparity in agricultural development.

In another study by Mathur, Das and Sarcer (2006) have identified the importance of public sector investment through credit facilities, insurance, and incentive measures. The study further insists upon the effect of agricultural price and fertilizer consumption to achieve agricultural growth. The role of government in agricultural development was also assessed by Kumar (2015). The development of agriculture lies in the availability of efficient infrastructure supported by the government.

The available literature on agricultural development subsumed the fact that institutional aspects such as land tenancy and distribution, irrigation infrastructure, technological implements, use of fertilizer and pesticides, nature of seeds, socioeconomic quality of farmers, government policy, and changing environment are fundamental determinants of it. In forthcoming section, variables and indicators are selected from abovementioned dimensions from best possible data sources to calculate composite index.

In the present section, variables and indicators have been selected based on proper literature survey. A total of 32 variables are categorized into 7 indicators based on previous studies. The list of indicators and noted variables is explained here under.

The first indicator is land. Land is the prime unit for cultivation. Here, percentage of land size holders under marginal (<1 ha), small (1–2 ha), semi-medium (2–4), medium (4–10 ha), large (> 10 ha), and all land size classes (Agriculture Census 2015–2016) has been taken into the list to know the nature of land size units in different blocks. Followed by this, average size of landholding is selected to acquire information on average land parcels. In developed countries, Agricultural practice in developed countries is very much encouraging because of both large size of landholding and bigger land parcel, and such agricultural systems provide an impetus for mechanized farming (Ghose and Saith 1976). In developing countries, study found that the location of cultivable land whether within village or outside village is an important dimension in decision-making to grow different kinds of crops. Following on the same line, own and self-operated land and lease in land are dimensions in decision-making regarding cultivation of crops. On land leased in land, marginal farmers prefer to grow cereals rather than cash crops due to their family consumption.

Irrigation is one of the most important indicators to analyze the agricultural situation of a region. If the percentage of area under assured irrigation is high, then different crops at a time can be grown which truly sensed better agricultural development (Awulachew et al. 2010). In climate change era, apart from assured irrigation infrastructure, the irrigation sources are also important aspect to channel water efficiently in the field. Tube well, tank, and sprinkle irrigation sources hence become important. Apart from these, the cost of irrigation also depends on whether irrigation technology is operated by diesel or electricity. Cost-effective assured irrigation infrastructure not only boosts diversification in agriculture, but it also increases gross cropped area.

One of the reasons for the success of the green revolution is the introduction of high yield varieties of seeds which boomed up productivity of food crops. The use of a higher percentage of HYV is good for the production system especially rural peasants of third world countries (Mundinamani 1985). The quality of seeds to be sown gets assurance from government agencies or institutions as certified seeds. Because of the higher cost of HYV seeds, the certified seed ensures other options to poor farmers in hope of good production.

Agricultural infrastructure requires money to buy. Marginal and small land size holders encourage more in a diversified production system if the government provides incentive such credit facilities at minimum interest rate. A handful study reveals that farmers avail credit from primary agricultural credit society (PACS) and regional rural banks (RRB) cultivate more profitable crops than those are not (Devi and Govt 2012). Therefore, assured credit facilities must be ensured in favors of agricultural innovation.

Mechanized farming is the need of the hour. Cost- and time-effective implements have increased work efficiency of farmers. Outsource income other than agriculture can be gained if small and marginal cultivators get mechanized farming opportunities. One of the advantages of developed agriculture remains in mechanized farming systems. The hand and animal operated implement in farming is the characteristic of subsistence agriculture, while the use of power operated machines suggests subsistence cum commercial or commercial agricultural systems. In subsistence farming, heavy machineries are not suitable as land size is small to marginal (Sen 1981). Therefore, power-operated small farm tractor, harvester, thresher, and hand operated implements are used.

As a supplement of natural soil nutrients, fertilizers and insecticides are used for desired soil quality. Next to HYV seeds and irrigation, fertilizer and pesticides were important green revolution infrastructure that attributed positively to higher productivity of cereals crops (Husain 1996). Per hectare fertilizer and pesticide inputs in developed agriculture is much higher than the average Indian agriculture. The common fertilizers used in rural India are urea, DAP, potassium, phosphate, and limited biofertilizers.

Socioeconomic variables such as average of farmers, nature work whether permanent or seasonal, literacy, and average family size are important attributes in the cultivation system. Adaptation of agricultural innovation depends upon the quality of farmers; hence literate peasants are more progressive than illiterates. In the farming system, the head of the households generally makes decisions about which type of agricultural system to be adopted. Study found that young literates are more adoptive toward agricultural innovation than aged (Asfaw and Admassie 2004). Agricultural innovation requires man power, and in the third world farming system, most of the work related to this was carried out by household's family; members therefore average family size is another dimension of agricultural development.

3.4   Trends and Dynamic Relation: Inputs-Outputs of Agriculture

The understanding of agricultural input-output relation is essential for improving agricultural growth and development. For the development of agriculture, input infrastructures are necessary. If this occurs, then the objective of intensive agriculture for the sake of food and livelihood could be achieved.

3.4.1   Trends in Inputs-Outputs of Agriculture

The essence of this section is to find out the input-output relation of agriculture in the district. The input study focuses on both physical and non-physical determinants (Siddiqui 2007). Soil nutrients, temperature, rainfall, and relative humidity have been taken as physical inputs of agriculture. On the other hand, land tenancy, irrigations, seed, credit facilities, fertilizer availability, and some socioeconomic aspects are considered as vital drivers of agricultural outputs.

Trends of agricultural inputs and outputs have been presented in Table 3.2 and 3.3 respectively. In Table 3.2, the standardized score of both physical and human indicators has been presented, and their composite standard score is also being calculated at quinquennial year gap because agriculture census data is only available after 5 years. Table 3.3 shows agriculture output, (productivity index and crop diversification index) on a 5 years' gap. Since there are only two agriculture outputs (crop diversification and productivity), therefore, the composite score is not calculated because of the nature of the analysis.

3.4.2   Dynamic Relations: Inputs-Outputs Indicator of Agriculture

To measure the impact of agricultural inputs on agricultural outputs, ordinary least square (OLS) is used. OLS is being the types of linear least square method used to estimate parameter through following formulation.

$$Y = \beta_0 + \Sigma_{j=1...p} \beta_j X_j + \varepsilon$$

where Y is the dependent variable, β_0 is the intercept of the model, X_j corresponds to the jth explanatory variable of the model ($j = 1$ to p), and e is the random error with expectation 0 and variance σ^2 (Sarkar 2013).The second hypothesis of the study has been tested on the impact of agricultural inputs to agricultural outputs (Table 3.4). Here, regression output shows that the variation in physical indicators of agriculture

Table 3.2 Trend in agricultural inputs in Malda District

Agricultural inputs

Year	Index value of physical indicators					
	Macro nutrients of soil	**Micro soil nutrients**	**Temperature**	**Rainfall**	**Humidity**	**CSS**
1995–1996	0.617	0.604	0.586	0.378	0.405	**0.518**
2000–2001	0.459	0.413	0.547	0.389	0.57	**0.476**
2005–2006	0.49	0.476	0.543	0.367	0.686	**0.512**
2010–2011	0.341	0.312	0.627	0.267	0.429	**0.395**
2015–2016	0.178	0.293	0.622	0.237	0.433	**0.352**
Year	**Index value of human indicators**					
	Irrigation	**Seeds**	**Credit system**	**Mechanization**	**Fertilizer and pesticides**	**CSS**
1995–1996	0.422	0.276	0.273	0.239	0.226	**0.287**
2000–2001	0.427	0.293	0.264	0.297	0.377	**0.331**
2005–2006	0.476	0.399	0.284	0.356	0.563	**0.416**
2010–2011	0.522	0.512	0.29	0.497	0.715	**0.507**
2015–2016	0.615	0.575	0.295	0.626	0.816	**0.586**

Source: Computed by author

Table 3.3 Trends in agricultural outputs in Malda District

Agricultural outputs

Year	Productivity (Yang's yield index)	Crop diversification index (Gibbs-Martin)
1995–1996	102.90	0.442
2000–2001	119.25	0.545
2005–2006	131.65	0.630
2010–2011	141.66	0.682
2015–2016	144.45	0.706

Source: Computed by author

Table 3.4 Regression summary (inputs-outputs)

Regression statistics	Productivity		Crop diversification	
	Physical inputs	Non-physical inputs	Physical inputs	Non-physical inputs
R square	0.644	0.902	0.618	0.897
Regression	758.03	1061.74	0.013	0.054
Residual	418.93	115.22	0.008	0.006
Total	1176.96	1176.96	0.022	0.06
P-value	0.01	0.001	0.014	0.004

Source: Computed by author

Table 3.5 Weight index for agriculture inputs-outputs

MDA summary		
Aspects	Physical index	Non-physical index
Weightage	50%	50%
Actual	0.315	0.449
Percent	32%	45%

Source: Computed by author

explains around 64% variability of productivity, while it is 90% in the case of the non-physical index. The CSS index of physical inputs explains approximately 62% of crop diversification, and it is 90% non-physical. Table 3.4 illustrates that non-physical indicators of agricultural output posit much better than the physical indicators. It also reports that non-physical index has a strong causal relationship with agricultural output in the district.

The null hypothesis whether physical and non-physical indicators has an equal impact on agricultural output or not is tested. Here, Table 3.5 based on a simple weighted index method shows that out of the same amount of weight (50% each for physical and non-physical index), the physical index explains around 32% of agricultural output and non-physical index describes 45%. It is, therefore, the non-physical index which finds more significant impact over physical indicators to agricultural outputs in the district. Thus, the null hypothesis fails to accept the proposition that both inputs equally impact on agricultural outputs.

3.5 Non-physical Inputs of Agriculture: Spatial-Temporal Dimensions

In this section, the composite condition of non-physical inputs of agriculture is presented by land size classes. Because of dearth of disaggregated data by land size classes at block level, physical aspects of agriculture have been dropped here.

The composite standardized score of 32 variables has been shown in Table 3.7. The composite normalized score indicates an increase under different land classes in district. In 1995–1996, the CSS was 0.350 which rose to 0.471 in 2015–2016. The pace of change in marginal land size category is higher which can be assessed through comparison between 1995–1996 and 2015–2016. In small land size class, it increased to 0.467 from 0.332 during the same two decades. The semi-medium category represents 0.357 in 1995–1996 which rose to 0.477 in 2015–2016 in district. The CSS in the medium class has changed from 0.327 to 0.477. Large land size category has recorded a change from 0.402 to 0.624 over 20 years in the district.

The CSS under all land size classes raised from 0.330 to 0.462. In the marginal land size class, five blocks have more than 0.400 index value among which Gazole records the maximum of 0.483 in 1995–1996. In the same year, Chanchal-II (0.463), Manikchak (0.462), Harishchandrapur-II (0.452), Harishchandrapur-I (0.417), and Chanchal (0.463) also remained in the list. During 1995–1996, the least scored block is Kaliachak-II with index value just 0.235 in marginal land size class in the district. Kaliachak-I (0.246), Kaliachak-III (0.270), Bamangola (0.273), and old Malda (0.281) blocks are following in the least category index. In 2015–2016, Gazole block remained at the top with index value 0.578 in the marginal category. In the same year, seven blocks have scored more than 0.500 index value in the district. Ratua-I (0.555), Chanchal- (0.550), Harishchandrapur-I (0.530), Manikchak (0.527), Harishchandrapur-II (0.507), and Chanchal-I (0.505) in the district are those blocks. In the same year, the least scored block is Kaliachak-II 0.344 in the marginal land size class group in the district. Old Malda (0.352), Kaliachak-III (0.3 54), and Kaliachak-I (0.432) have also scored low in the district (Table 3.6).

In 1995–1996 under small land size category, the highest score shows in Gazole block (0.495) and subsequently Harishchandrapur-II (0.473), Habibpur (0.425), Chanchal-II (0.365), Chanchal-I (0.349), Kaliachak-II (0.213), Kaliachak- (0.224), and Kaliachak-III (0.264) block in the district. In 2015–2016, Habibpur block locates maximum index among the blocks with index value 0.626. Except Habibpur block, there are other four blocks with scores more than 0.500 index value, namely, Chanchal-II (0.609), Gazole (0.607), Ratua-I (0.518), and Manikchak (0.506) in the district. In the same year, the lowest scored blocks are Kaliachak-III (0.307), Kaliachak-I (0.356), and old Malda (0.360) in the district.

Gazole with 0.505 CSS index is ranked first among the blocks in the district under semi-medium land size group in 1995–1996. In the same year, Ratua-I (0.413), Harishchandrapur-II (0.398), Habibpur (0.394), and Harishchandrapur-I (0.391) were in descending order in the list. The lowest score secured was Kaliachak-II with index value 0.255 followed by Old Malda (0.281) and Manikchak (0.304). With CSS index value 0.648, Habibpur block sits topped among in semi-medium class in 2015–2016 in the district. In other blocks, the index value reports more than 0.500 mainly in Gazole (0.613), Chanchal-II (0.603), Harishchandrapur-II (0.556), Harishchandrapur-I (0.517), and English Bazar (0.501) block, the lowest score appears in Manikchak (0.330), Old Malda (0.348), Kaliachak-II (0.386), and Kaliachak-I (0.399) in the district.

Table 3.6 List of non-physical indicators and variables

Indicators	Variables	Description of the variables
Land	**X1**	Percentage of land size holders
	X2	Average size of landholding
	X3	Percentage of area wholly own and self-operated
	X4	Percentage of leased area
	X5	Percentage of net sown area
	X6	Percentage of the area entirely within the village
	X7	Percentage of area outside of village
Irrigation	**X8**	Percentage of net irrigated area
	X9	Percentage of area irrigated by tank and other sources
	X10	Number of electric well and tube well in 10 ha of GCA
	X11	Number of diesel operated well and tube well in 10 ha of GCA
	X12	Percentage of all crops irrigated in total GCA
	X13	Percentage of all crops unirrigated in total GCA
Seeds	**X14**	Percentage of households used certified seeds
	X15	Percentage of households used HYV seeds
Credits	**X16**	Percentage of households avail institutional credit facilities
	X17	Percentage of households taken credit from PACS
	X18	Percentage of households taken credit from PLDB
	X19	Percentage of households taken credit from CBB
	X20	Percentage of households taken credit from RRB
Machineries	**X21**	Number of hand-operated machines and implements in per hectare of GCA
	X22	Number of animals operated machines and implements in per hectare of GCA
	X23	Number of power operated machines and implements in per hectare of GCA
Fertilizer and pesticides	**X24**	Percentage of GCA treated with fertilizer
	X25	Percentage of GCA treated with pesticides
	X26	Amount of NPK mixture used kg/hectare
	X27	Amount of urea, DAP, superphosphate mixture used kg/hectare
Socioeconomic	**X28**	Percentage of main worker in the total cultivator
	X29	Percentage of marginal worker in the total cultivator
	X30	Literacy rate
	X31	Average age
	X32	Average family size

Source: Based on literature survey by author
Note: data of mentioned variables listed in annexures 3f and 3g for 1995–1996 and 2015–2016, respectively

In the medium land size category, Gazole scored 0.468 which is the maximum among blocks in the district in 1995–1996. The blocks like Ratua-I (0.442), Habibpur (0.372), English Bazar (0.369), Bamangola (0.331), and Manikchak (0.329) are succeeding blocks in the list. The least CSS finds in Kaliachak-II (0.269), Old Malda (0.279) Harishchandrapur-I (0.281), and Chanchal-II (0.284). In 2015–2016, Habibpur scored a maximum that is 0.646. Ratua-I (0.553), Gazole (0.490), Bamangola (0.484), and Old Malda (0.448) have enjoyed better development in the district under said category. The least CSS is found in Manikchak (0.372), Kaliachak-III (0.385), Harishchandrapur-I (0.404), and Kaliachak-I (0.415) blocks in the district.

In 1995–1996 under large land size class category, Chanchal-I, Chanchal-II, Kaliachak-II, Manikchak, and Ratua-I were five such blocks where the large farmer had not been found in the district. The maximum index value was in English Bazar 0.526, block followed by Kaliachak -III (0.454), Gazole (0.430), Bamangola (0.427), and Ratua-I (0.425) blocks in the district. The lowest CSS was found in Harishchandrapur-I (0.162), Harishchandrapur-II (0.370), and Kaliachak-I (0.371) in the district. In 2015–2016, there are only two blocks where large farmers mainly found in Bamangola (0.685) and Kaliachak-III (0.564).

In 1995–1996, all land size class category has recorded the highest index value for Gazole block with 0.495 in the district. Other five successive blocks were Manikchak (0.423), Harishchandrapur-II (0.409), Habibpur (0.385), Chanchal-II (0.351), and Ratua-I (0.344) in the district. In the same year, the least scored block was Kaliachak-II (0.217) followed by Kaliachak-I (0.227), Kaliachak-III (0.285), and Old Malda (0.288). In 2015–2016, Gazole was the most developed block (0.587) in the district. The other blocks like Chanchal-II (0.576), Habibpur (0.571), Ratua-I (0.533), and Manikchak (0.493) have also scored high index value in the district. The last score records in Kaliachak-II (0.326) followed by Kaliachak-III (0.339), Old Malda (0.348), and Kaliachak-I (0.375) in the district.

Land size class category wise change in CSS index has been presented in Fig. 3.2 based on Table 3.7. The maximum change has been recorded under large land size class category with index value 0.222 followed by small (0.135), all classes (0.131), marginal (0.121), and semi-medium (0.120), and it was lowest under medium land size group in descending order in district. Among blocks, the maximum amount of change in CCS index is located under medium land size group in Habibpur block with index value 0.274 in the district. The minimum change has been recorded in Harishchandrapur-II under small land size type, i.e., 0.014 in the district.

The block level analysis denotes that the nature of change under marginal land size class is persistent throughout the blocks. The change is maximum in said category as 0.244 in Ratua-I followed by Kaliachak-I (0.185) and Ratua-II (0.168) blocks suggest higher pace of development. The least change in CSS value is found under Harishchandrapur-II (0.055), Manikchak (0.066), and Old Malda (0.071) indicating the slow pace of development in agricultural inputs in the district. In the small land size category, maximum change is recorded under Chanchal-II with index

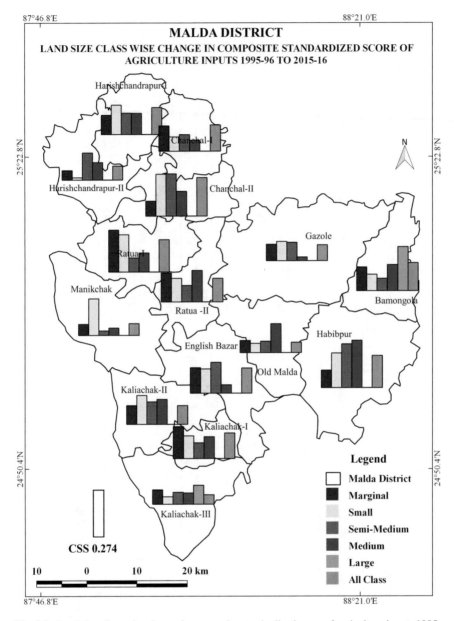

Fig. 3.2 Land size class wise change in composite standardized score of agriculture inputs 1995–1996 to 2015–2016. (Source: based on Table 3.7)

Table 3.7 Composite standardized score of agriculture inputs by land size class in Malda District, 1995–1996 and 2015–2016

Block	1995–1996						2015–2016					
	Marginal	Small	Semi-medium	Medium	Large	All classes	Marginal	Small	Semi-medium	Medium	Large	All classes
Bamangola												
Chanchal-I	0.358	0.349	0.348	0.306	N. A	0.459	0.505	0.432	0.446	0.370	N. A	0.307
Chanchal-II	0.463	0.365	0.356	0.284	N. A	0.576	0.550	0.609	0.603	0.429	N. A	0.351
English Bazar	0.318	0.307	0.322	0.369	0.562	0.446	0.466	0.448	0.501	0.418	N. A	0.300
Gazole	0.483	0.495	0.505	0.468	0.430	0.587	0.578	0.607	0.613	0.490	N. A	0.495
Habibpur	0.363	0.425	0.394	0.372	0.418	0.571	0.465	0.626	0.648	0.646	N. A	0.385
Harishchandrapur-I	0.417	0.305	0.391	0.281	0.162	0.471	0.530	0.477	0.517	0.404	N. A	0.313
Harishchandrapur-II	0.452	0.473	0.398	0.314	0.370	0.490	0.507	0.487	0.556	0.417	N. A	0.409
Kaliachak-I	0.246	0.224	0.308	0.289	0.371	0.375	0.432	0.356	0.399	0.415	N. A	0.227
Kaliachak-II	0.235	0.213	0.255	0.269	N. A	0.326	0.344	0.383	0.386	0.415	N. A	0.217
Kaliachak-III	0.270	0.264	0.349	0.319	0.454	0.339	0.354	0.307	0.419	0.385	0.564	0.285
Manikchak	0.462	0.294	0.304	0.329	N. A	0.493	0.527	0.506	0.330	0.372	N. A	0.423
Old Malda	0.281	0.305	0.281	0.279	0.381	0.348	0.352	0.360	0.348	0.448	N. A	0.288
Ratua-I	0.311	0.301	0.413	0.442	0.425	0.533	0.555	0.518	0.496	0.553	N. A	0.344
Ratua-II	0.318	0.328	0.360	0.261	N. A	0.458	0.486	0.465	0.455	0.443	N. A	0.321
District	0.350	0.332	0.357	0.327	0.402	0.462	0.471	0.467	0.477	0.446	0.624	0.330

Source: Computed by author (annexure 3f and 3g)

N. A not available

value 0.244 and least change in Harishchandrapur-II (0.014). In the semi-medium land size class category, the maximum difference appears in Habibpur (0.254) and Chanchal-II (0.246) blocks. Manikchak block witnesses a minimum change with index value 0.026 in said land size class category. In medium class, the maximum amount of change reports in Habibpur (0.274), while the least difference sits in Gazole (0.022) block. Since there are only two blocks in a large land size class group, therefore, the change in index value is only calculated for Bamangola (0.257) which is higher than Kaliachak-III (0.110) block. Under all land size class groups, the maximum change in index value is found for Chanchal-II, i.e., 0.225 followed by Ratua-II (0.189). The minimum difference is reported in Old Malda (0.060).

3.6   Agricultural Outputs

Agricultural output is considered as the total outcome of a farm against total inputs in a year. The output dimensions of agriculture (whether total production or monetary account) is reasonably expressed through cropping patterns, productivity, and diversity of crops that added in total output (OECD 1999). Among these three output dimensions, cropping intensity is not direct agricultural output because it reflects in areal extent of crops which does not guarantee better output; therefore, cropping intensity is considered as indirect output measure of agricultural development (Siddiqui et al. 2011), and therefore, in present analysis, cropping intensity as output dimension is dropped. The monetary value of total production is added value from different crops; therefore, crop diversification and productivity are direct and essential outputs of agriculture.

3.6.1   Cropping Intensity

Cropping intensity is defined as a ratio between gross cropped area (GCA) and net sown area (NSA) in a region at a point of time. It thus indicates the additional percentage share of the area sown more than once to NSA. It shows that the extent to which the productive capacity of land is being utilized is supported by modern technological and infrastructures. The formula to calculate cropping intensity is

$$\text{Cropping Intensity} = \frac{\text{Gross Cropped Area}}{\text{Net Sown Area}} \times 100$$

In general, the level of cropping intensity is higher in the regions with higher percentage of net sown area and with higher intensity of land use by modern technologies. This technique is used in Chap. 5.

Table 3.8 Method to calculate crop yield index (W.Y Yang's 1965)

Name of the crops	The area under crop in the block (area in hectare)	Yield in quintal/ hectare		Crop yield in the block as the percentage to the district	Percentage multiply by area (in hectares)
		Average yield in district	Average yield in the block		
1	2	3	4	5 = Col. 3/ col4*100	6 = col.5*col. 2
Cereals	21,541	5122	2800	182.93	3940464.36
Pulses	6	1124	1298	86.59	519.57
Spices	2	1423	990	143.74	287.47
Fruits	5	3965	2413	164.32	821.59
Vegetables	3255	4562	2791	163.45	532042.64
Oilseeds	3050	851	963	88.37	269527.52
Fibers	5	5632	3892	144.71	723.54
Total	27,864				4744386.68
Computation of crop yield index for Bamangola block marginal class, 2015–2016					**4744386.68/27864**
					170.27

Source: Calculated by author

3.6.2 Productivity

Productivity is defined merely as the input-output ratio in any production system. Here inputs include both physical and non-physical, and output is the total production. Productivity is a broader concept, and it has a multidisciplinary dimension in different discipline. Therefore, there are many methods and technique to compute agricultural productivity. The present study uses W.Y. Yang (1965) methods to compute productivity because of data convenience. He has used "Crop Yield Index" for calculating agricultural productivity. In Table 3.8, first column top seven crops categories are listed. In column two, the actual area in hectares occupied by each crop category is noted. In column three and four, average yield which is calculated from total production divided by total area under specific crop is presented at district and block level. The average yield columns show that except pulses and oilseeds, other crops report better production in district than block. In column five, the percentage of block yield on district is computed, and in column six, percentage of block yield is multiplied by block area. In the last row of column six, Yang's yield index is computed by dividing total of column six to total of second column. In Bamangola block under marginal land size group, Yang's yield index is 170.27 in 2015–2016.

There is a significant increase in productivity index under different land size classes in the district from 1995–1996 to 2015–2016. The change in productivity in marginal land size class category is the maximum, i.e., 48.87, which signifies marginal farmers practice most intensive cultivation as compared to the rest of the land size groups. In the district, productivity has increased under small (43.40), semi-medium (44.64), medium (43.49), large (23.38), and in all land size class (43.76). Among the block, the maximum increase has been recorded under all land size class group in Kaliachak-I (58.89) block. The minimum increase is noted under large land size class in Bamangola block (11.88).

3.6.3 Crop Diversification

Crop diversification refers to the growth of multiple crops in a region at a time. The diversified cropping pattern of a region is induced by advanced agricultural innovations supported by institutional, infrastructural, technological, and socioeconomic facilities (Singh 1976). The crop diversification ensues when less remunerative crops replace high value crops because it offers better income opportunities in the existing agricultural system. And that's why crop diversification is considered as an important output dimension of agricultural development of a region. The recent experience in Indian agriculture is increasingly focused on crop diversification through policy interference at macro and micro level to boost regional development.

   In academic discourse, crop diversification is a multidimensional concept. In different disciplinary perspectives such as biology, economics, management, geography, sociology, and other domains of knowledge, the meaning of diversification differs. And therefore, to measure crop diversification and its intensity at a different level, there are many methods such as Gibbs and Martin's, Bhatia and Singh, Simpson diversification index, and others. In the present study, Gibbs-Martin (1962) crop diversification index has been used because his technique measures diversification up to an aerial extent of crops that occupy up to 0.1% in the total cropped area of a region (district/block). His diversification index considers the percentage of all crops of the total cropped area.

$$\text{Gibbs and Martin's Technique} = 1 - \frac{\Sigma X^2}{\left(\Sigma X\right)^2}$$

Here, X is the percentage of area occupied by an individual crop at a point of time. The index value ranges between 0 and 1. The index value approach to 1 indicates increasing diversification and toward 0 suggests lower diversification. This technique helps us understand the magnitude of diversification. The same technique is used in Chaps. 4 and 5 also.

   In the study it is found that there is a positive change in crop diversification index across different land size class categories in the district from 1995–1996 to 2015–2016. Marginal (0.094), small (0.167), semi-medium (0.074), large (0.574), and all land size classes (0.263) locate the positive change in district. At the block level, maximum positive change (0.464) is found under all land size class category in Ratua-II, while Manikchak notes maximum negative change (0.522) under medium land size class in the district. Except two blocks (Harishchandrapur-II and Old Malda) under marginal land size class and three blocks in small land size category, the remaining blocks report positive change in crop diversification index.

3.7 Agricultural Outputs: Spatial-Temporal Dimensions

In this section, the change in agriculture development based on outputs indicators (productivity and crop diversification) is presented. The CSS of outputs is presented in Table 3.9, and change in CSS index from 1995–1996 to 2015–2016 is displayed in Fig. 3.3.

The pace of change in agricultural outputs seems higher in marginal land category due to higher non-physical inputs. As it has been proved that the better non-physical inputs lead to higher crop diversification, therefore spatial variations in change of agricultural outputs have anticipated relation with non-physical inputs. The other land size categories have also recorded an increase, but tempo of change is slower than marginal land size class. The small and semi-medium land size classes have found substantial improvement due to a positive and skewed productivity curve, but crop diversification is recorded at a slower pace of improvement than productivity.

Figure 3.3 shows that under different land categories, the index value is increased among them and the large land size class records maximum positive change with index value 0.378. In marginal (0.082), small (0.090), semi-medium (0.094), medium (0.040), and all land size class (0.077) appear the positive growth due to increase in productivity and crop diversification in the district. Among the blocks, the maximum gain reports under medium land size category in Old Malda (0.316), while minimum increase is found under medium land category in Chanchal-I (0.004) block.

The overall scenario of agricultural outputs for Bamangola, Habibpur, Chanchal-II, and Harishchandrapur-I shows positive change due to better share of diversification index. The non-physical input is also found as an enhancement in different blocks.

3.8 Karl Pearson's Correlation of Coefficients (Inputs and Outputs)

Karl Pearson's coefficient of correlation is a statistical technique used to measure the degree of the linear association between variables. The degree of associations between variables is determined in -1.00 to 1.00 scale range which is known as coefficient of correlation. If the calculated value is near to -1.00, then relation among variables is strongly negative meaning; thereby the increase in one variable causes the other. Similarly, correlation of coefficient nearer to 1.00 stands as positive

Table 3.9 Composite index of agriculture outputs by land size class in Malda District 1995–1996 and 2015–2016

Block	1995–1996						2015–2016					
	Marginal	Small	Semi-medium	Medium	Large	All classes	Marginal	Small	Semi-medium	Medium	Large	All classes
Bamangola	0.406	0.503	0.279	0.393	0.232	0.400	0.467	0.550	0.426	0.582	0.371	0.454
Chanchal-I	0.691	0.710	0.665	0.575	N. A	0.698	0.730	0.784	0.789	0.579	N. A	0.799
Chanchal-II	0.574	0.569	0.587	0.593	N. A	0.585	0.688	0.717	0.693	0.745	N. A	0.696
English Bazar	0.618	0.668	0.568	0.468	0.346	0.621	0.665	0.691	0.650	0.586	N. A	0.664
Gazole	0.370	0.337	0.250	0.275	0.333	0.340	0.440	0.430	0.328	0.364	N. A	0.433
Habibpur	0.248	0.296	0.320	0.279	0.377	0.319	0.423	0.476	0.504	0.441	N. A	0.429
Harishchandrapur-I	0.592	0.511	0.310	0.484	0.284	0.502	0.659	0.659	0.489	0.500	N. A	0.650
Harishchandrapur-II	0.739	0.712	0.735	0.784	0.482	0.733	0.786	0.770	0.797	0.798	N. A	0.798
Kaliachak-I	0.301	0.339	0.440	0.256	0.032	0.354	0.373	0.452	0.466	0.418	N. A	0.409
Kaliachak-II	0.357	0.363	0.306	0.455	N. A	0.417	0.451	0.433	0.395	0.499	N. A	0.432
Kaliachak-III	0.731	0.751	0.657	0.732	0.785	0.743	0.797	0.797	0.730	0.769	0.812	0.786
Manikchak	0.741	0.747	0.857	0.629	N. A	0.811	0.868	0.837	0.883	0.766	N. A	0.852
Old Malda	0.426	0.407	0.265	0.334	0.302	0.382	0.505	0.496	0.288	0.650	N. A	0.483
Ratua-I	0.744	0.676	0.836	0.690	0.445	0.800	0.809	0.750	0.887	0.791	N. A	0.830
Ratua-II	0.578	0.627	0.527	0.617	N. A	0.529	0.686	0.725	0.682	0.687	N. A	0.700
District	0.541	0.548	0.507	0.544	0.372	0.549	0.623	0.638	0.600	0.584	0.750	0.626

Source: Computed by author (see annexure 3h)

N. A not available

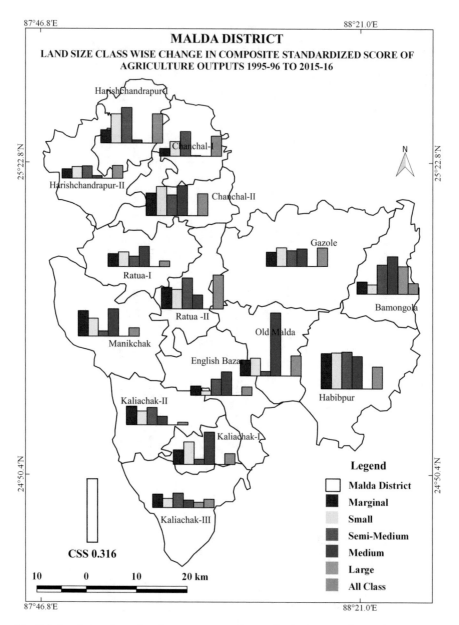

Fig. 3.3 Land size class wise change in composite standardized score of agricultural outputs 1995–1996 to 2015–2016. (Source: based on Table 3.9)

and strong relation and away from it posits low or weak relation among correlated variables. Karl Pearson's equation is given below:

$$r = \frac{\sum (X - \underline{X})(Y - \underline{Y})}{\sqrt{\left(X - (X)^2\right)}\sqrt{(Y - \underline{Y})^2}}$$

where $\underline{X}$ = mean of X variable
     $\underline{Y}$ = mean of the Y variable

Here an attempt has been made to calculate the degree of relation between agricultural inputs (seven indicators) and agricultural outputs (two) in Malda District. The correlation of coefficients is the reason behind choosing indicators (inputs-outputs) to measure agriculture development. The strong positive or negative relation with inputs and output indicators will not only justify the validity of the selection of indicators, but it also illustrates the degree of associations between indicators.

3.8.1   Inputs-Outputs Relation of Agriculture Indicators

Table 3.10 shows that the output indicator of agriculture, i.e., crop diversification, is positively correlated with irrigation, machinery, and fertilizers in marginal land size class category in 1995–1996. Except the first input indicator (land), other indicators are related positively with crop diversification across land size categories. The correlation coefficient between output indicators (crop diversification and productivity) and input indicators like land, irrigation, seeds, fertilizers, and socioeconomic indicators become more robust in 2015–2016 under the said land size category.

In small land size class group, the input indicators such as irrigation, seeds, machinery, and fertilizers report positive and strong association with crop diversification in 1995–1996. Again, the relationship becomes robust in 2015–2016 with output indicator such as crop diversification. The productivity also shows positive and robust relation with land, irrigation, and seed.

Under semi-medium land size category, the output indicators such as crop diversification and productivity reveal a positive association with irrigation and machinery, and in case of productivity, land, machinery, and fertilizer consumption display positive relation which becomes stronger in 2015–2016.

In 1995–1996, the selected indicator exhibits positive relation with crop diversification and productivity under medium land size class in the district. The input indicators like machinery and fertilizers note strong and positive correlation with diversification, while irrigation and seeds are significantly and positively

Table 3.10 Karl Pearson's correlation of coefficient of agricultural inputs and output in Malda District 1995–1996 and 2015–2016

Indicators	Crop diversification											
	1995–1996						2015–2016					
	Marginal	Small	Semi-medium	Medium	Large	All class	Marginal	Small	Semi-medium	Medium	Large	All class
X₁	−0.351	−0.375	0.347	0.353	0.244	−0.481	−0.175	0.613*	0.436	−0.154	0.966***	−0.444
X₂	0.714**	0.668**	0.536*	0.459	0.535*	0.734**	0.713**	0.775***	0.501*	0.998***	0.896***	0.857***
X₃	0.425	0.596*	−0.265	0.315	0.498*	0.799***	0.596	0.521*	−0.181	0.487	0.964***	0.726**
X₄	0.388	0.163	−0.124	0.096	0.021	0.595*	0.321	−0.361	−0.245	−0.236	0.570*	0.473
X₅	0.684**	0.524*	0.669**	0.558*	0.526*	0.709**	0.531*	0.461	0.512*	0.698**	0.804***	0.818***
X₆	0.582*	0.501*	0.447	0.591*	0.530*	0.893***	0.829***	0.398	0.881***	0.756**	0.725**	0.799***
X₇	0.449	−0.419	−0.27	0.264	0.535*	0.398	0.522*	0.599*	0.772***	0.441	0.951***	0.482
Productivity												
X₁	0.741**	0.779**	0.832***	0.621*	0.979***	0.441	0.841***	0.997***	0.921***	0.621*	0.979***	0.667**
X₂	0.517*	0.920***	0.612*	0.822***	0.833***	0.696**	0.887***	0.832***	1.000***	0.716**	0.863***	0.772***
X₃	0.615*	0.706**	0.261	0.813***	0.792***	0.614*	0.596*	0.399	−0.324	0.492*	0.772***	0.774***
X₄	0.068	0.29	0.007	0.491*	−0.175	0.189	0.519*	−0.412	0.436	0.251	0.501*	0.553*
X₅	0.790***	0.304	0.801***	0.441	0.875***	0.587*	0.812***	−0.043	0.742**	0.776***	0.875***	0.711**
X₆	0.750**	0.449	0.668**	0.512*	0.857***	0.774***	0.623*	0.732**	0.853***	1.000***	0.567*	0.825***
X₇	−0.359	0.145	0.228	0.398	0.822***	−0.026	0.278	0.666**	−0.361	−0.339	−0.402	0.576*

Source: Calculated by author

Inputs – X_1, land; X_2, irrigation; X_3, seeds; X_4, credits; X_5, machinery; X_6, fertilizers and pesticides; X_7, socioeconomic

Outputs – Crop diversification and productivity

Note. $* p < 0.05$, $** p < 0.01$, $*** p < 0.001$

related with productivity. In 2015–2016, the degree of association with input indicators became robust with crop diversification and productivity.

The more robust and positive association is noted under the large land size category with crop diversification and productivity in 2015–2016 as compared to 1995–1996 in the district. The input indicators such as irrigation, seed, machinery, fertilizer, and socioeconomic display positive correlation with diversification in 1995–1996, and except credit, all other selected variables are found to be strong and having positive correlation with productivity in the same year in district.

Under all land size class category, excluding land, all other input indicators result in positive correlation with crop diversification and except socioeconomic and other indicators found positive association with productivity in 1995–1996. The amount of correlation with selected indicators became stronger in 2015–2016 over 1995–1996 with crop diversification and productivity as well in district.

3.8.2   Non-physical Inputs and Crop Diversification

The overall relationship between crop diversification and non-physical indicators has been presented in Table 3.11. This computation aims to know the kind of relationships that have existed between crop diversification and non-physical inputs of agriculture. The third null hypothesis of the study has claimed that "there is a negative correlation between crop diversification and non-physical inputs of agriculture" in district. Table 3.11 shows that there is and remains a positive linear association between crop diversification and non-physical inputs of agriculture in 1995–1996 and 2015–2016. Hence, it can be said that if non-physical inputs of agriculture increase over time, then crop diversification will tend to increase.

Table 3.11 explains that the degree of Karl Pearson correlation coefficient in 1995–1996 stood as 0.603 which became more robust in 2015–2016 (0.773). The Karl Pearson's correlation of coefficient between crop diversification and non-physical inputs of agriculture is significant at 0.05 level in 1995–1996 and improved to 0.000 in 2015–2016.

Table 3.11   Karl Pearson's correlation of coefficient

	Non-physical agricultural inputs	
	1995–1996	2015–2016
Crop diversification	0.603*	0.773***
Sig. (2-tailed)	0.018	0.000
N	85	77

Source: Calculated by author
*Correlation is significant at 0.05 significance level (2-tailed)
***Correlation is significant at the 0.001 significance level (2-tailed)

3.9   Spatial-Temporal Pattern of Agricultural Development: Inputs-Output Dimensions

Available literature on agricultural development reveals that the measurement process of it incorporates both types of variables (inputs-outputs) to assess spatial and temporal change in agricultural development of a region. It is necessary because agricultural development means overall positive change in agricultural processes irrespective of inputs or output dimensions (Nath 1969; Alam 1974; Ram 1989; Bhat 2013 cited Raza 1980).

In this section, the status of agricultural development is presented land size category wise for 1995–1996 and 2015–2016 at block level. To show spatial patterns of change in development index, the quartile technique has been used. The composite score of agricultural inputs and outputs has been categorized into three classes where first, second, and third quartiles are taken as breaking points of each index category (Table 3.12). The first breaking point of composite inputs data series of 1995–1996 is 0.341 as lower quartile and 0.503 as upper quartile, and the same became 0.416 and 0.607 in 2015–2016, respectively.

To show the overall picture of agriculture development (inputs and output) at the block level, the calculated CSS is presented in spatial maps (Figs. 3.5, 3.6, 3.7, 3.8, 3.9). Because of an increase in composite standardized score from 1995–1996 to 2015–2016 under different land size classes, the common breaking point is not preserved in both the years. It is found that the blocks lying above the upper quartile index have a familiar pattern of higher inputs and better output. Similarly, lower quartiles led us to the conclusion that lower inputs lead lower outputs. The medium class is enjoying a win-win situation in terms of inputs and outputs of agriculture in both the years.

It is seen from Table 3.10 that the degree of correlation between inputs and outputs of agriculture under different land size categories in 1995–1996 and 2015–2016 is mostly positive. It is clear that agricultural inputs like irrigation, fertilizers, seeds, and machinery are strong and positively correlated with crop diversification and productivity where land, socioeconomic, and credit show moderate to high and in some cases lower degree of association with agricultural outputs in the district. If strongly correlated input indicators have increased, then the output of agriculture definitely would increase. An attempt has been made to show that the spatial pattern and variations of such input induces output agriculture at the block level.

Table 3.12 Determination of breaking points to agriculture development

Quartile division	1995–1996	2015–2016
Upper quartile	0.503	0.607
Lower quartile	0.341	0.416

Source: Calculated by author

3.9.1   Agriculture Developments in Marginal Land Size Category

Figure 3.3 illustrates the agriculture development in marginal land size class for 1995–1996 and 2015–2016 in the district. In 1995–1996, six blocks were having index value above the upper quartile (> 0.503) and the blocks were Harishchandrapur-I, II, Chanchal-I, II, Ratua-I, and Manikchak. Except Harishchandrapur-II, other blocks retained their position over and above the upper quartile score (>0.607) in 2015–2016. The better agricultural innovation such as insured irrigation, HYV seeds, chemical fertilizers, intensive farming, and quality of farmers have been important causes to support higher agricultural development in these noted blocks. In 1995–1996 and 2015–2016, Manikchak block remains the most developed block with index value 0.602 and 0.698, respectively. The index value between 0.341 and 0.503 is reported in five blocks (Ratua-II, Gazole, Old Malda, English Bazar, and Kaliachak-III), among them Kaliachak-III (0.500) is the most developed block. In 2015–2016, the same five blocks and other two blocks (Bamangola and Habibpur) were found in medium development (0.416 to 0.607) range. In the low agriculture development range, four blocks namely Kaliachak-I, II, Habibpur and Bamangola were placed in 1995–1996, but only two blocks remain under lower quartile range (<0.416) in the district in 2015–2016 (Fig. 3.4).

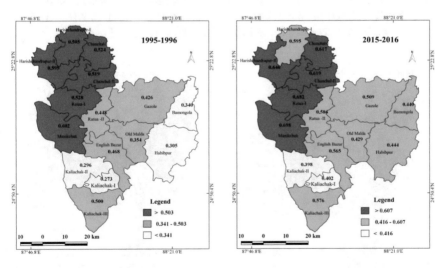

Fig. 3.4 Agricultural development (inputs-outputs) under marginal land size category in Malda District 1995–1996 and 2015–2016

3.9.2 Agriculture Developments in Small Land Size Category

In 1995–96, four blocks were reported above upper quartile index (>0.503) namely Harishchandrapur-II, Chanchal-I, Kaliachak-III and Manikchak in the district. The modern agricultural infrastructure such as use of fertilizers, pesticides, HYV seeds, and better socioeconomic conditions of the farmers are major forts that remain behind higher development. In 2015–2016 except Kaliachak-III, other blocks remained in high agricultural development category (>0.607), while Chanchal-II and Ratua-I transformed from medium to high development range. The low agricultural development (<0.341) is found only in two blocks (Kaliachak-I and II) and also remains below lower quartile (<0.416) range in 2015–2016. In 1995–1996, there were nine blocks found in medium development range, and in 2015–2016, the same development range shows eight blocks with wide transformation (Fig. 3.5).

3.9.3 Agriculture Developments in Semi-Medium Land Size Category

The modern agricultural infrastructure such as fertilizer, machinery, and HYV seeds leads to higher productivity thus higher agriculture development in semi-medium land size category. In 1995–1996, there were four blocks in high development range (>0.503) while in 2015–2016, though the number of blocks remained the same but Chanchal-II replaced Manikchak in the same category (>0.607). In 2015–2016 and 1995–1996, the number of blocks in medium range remained same, but the only

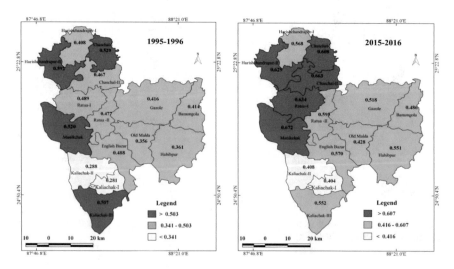

Fig. 3.5 Agricultural development (inputs-outputs) under small land size category in Malda District 1995–1996 and 2015–2016

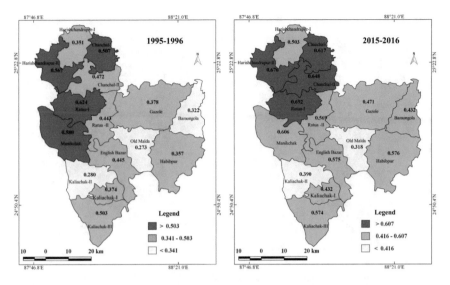

Fig. 3.6 Agricultural development (inputs-outputs) under semi-medium land size category in Malda District 1995–1996 and 2015–2016

change was noticed in case of Bamangola block which was positioned below lower quartile (<0.341) in 1995–1996 and now upgraded into medium agriculture development range. In 1995–1996, there were three blocks (Bamangola, Old Malda, and Kaliachak-II) reported in the low development category, and in 2015–2016, the same low development index was noted in two blocks and Bamangola block upgraded to medium development range (Fig. 3.6).

3.9.4   Agriculture Developments in Medium Land Size Class

The agriculture development under medium land size category in 1995–1996 shows that three blocks, viz., Harishchandrapur-II, Ratua-I, and Kaliachak-III appear above the upper quartile (> 0.503) induced by high fertilizer consumption, extended irrigation facilities, use of HYV seeds and intensive farming. In 2015–2016, though the number of blocks remained the same, Manikchak block was transformed from medium to high development range. In 1995–1996, there were nine blocks exhibiting medium development, and in 2015–2016, the number of blocks remains the same (Fig. 3.7). The transformation shows in Old Malda and Habibpur blocks from lower to medium development range. In low agricultural development range, the number of blocks is three in both years. The Kaliachak-I block only remains in the same development range, but Old Malda and Habibpur have replaced Bamangola and Gazole in 2015–2016 which were in medium development range in 1995–1995.

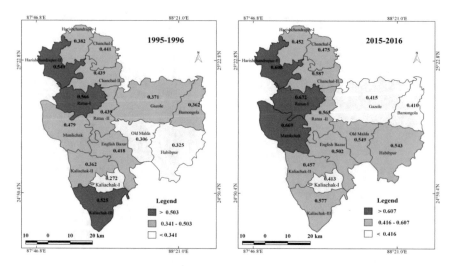

Fig. 3.7 Agricultural development (inputs-outputs) under medium land size category in Malda District 1995–1996 and 2015–2016

3.9.5 Agriculture Developments in Large Land Size Class

In 1995–1996, there were five blocks (Chanchal-I, II, Manikchak, Kaliachak-II, and Ratua-II) where large land size farmers were not reported. Due to high population density and increasing nucleated family size, the number of large farmers in the district is very small. Among blocks where large farmers are ensured, Kaliachak-III block is sited in high agricultural development range and four blocks, namely, Harishchandrapur-I, Kaliachak-I, Old Malda, and Bamangola, which exhibit low agricultural development. The low agricultural development is mainly due to low productivity and diversification. The other four blocks were located in medium agricultural development range. In the 2015–2016 Census, there are 13 blocks where no large farmers are reported. In the same Census, there is only one block (Bamangola) depicted in medium development range, and Kaliachak-III is reported in high agricultural development range (Fig. 3.8). The marginalization of land size because of family nucleation and increasing population has been decreasing land size in the district.

3.9.6 Agriculture Developments in All Land Size Classes

Figure 3.9 illustrates that in 1995–1996, five blocks, namely, Chanchal-I, Harishchandrapur-II, Ratua-I, Manikchak, and Kaliachak-III, were in high agriculture development range, and among them, Manikchak sites at top, i.e., 0.617

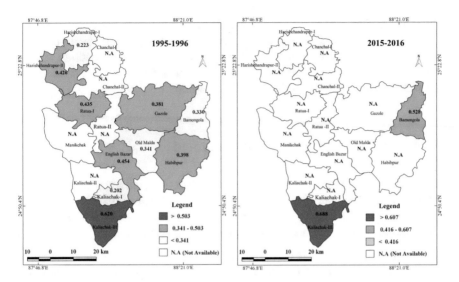

Fig. 3.8 Agricultural development (inputs-outputs) under large land size category in Malda District 1995–1996 and 2015–2016

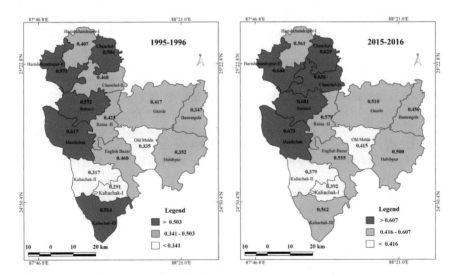

Fig. 3.9 Agricultural development (inputs-outputs) under all class land size category in Malda District 1995–1996 and 2015–2016

index value. These mentioned blocks have reported better productivity led by HYV seeds, fertilizer, irrigation, mechanized farming, and innovative agriculture practices. In 2015–2016, although six blocks are located in development range (0.607), Chanchal-II is located into high development category, while Kaliachak-III comes down to medium development range. In 1995–1996, seven and three blocks were found in medium and low development index, and in 2015–2016, the number of blocks in both categories remains the same. The transformation in the development landscape is reported to Kaliachak-III block which is placed in medium range in 2015–2016 from high in 1995–1996. In 2015–2016, Chanchal-II promotes in to upper quartile range which indicates better development pace.

3.10 Pace of Agricultural Development

Spatial analyses on agricultural development suggest overall improvement in agricultural saturation in the district. However, the pace of development under land size classes is not same rather differs significantly. The nature and pace of change in agriculture development land scape show following characteristics:

1. Under marginal land size class, the maximum pace of change is noted in Ratua-II block which added 0.154 index value in two decades. Followed by this, Habibpur (0.139), Ratua-II (0.138), Kaliachak-I (0.129) and Kaliachak-II, Bamangola, and Chanchal-II have added above 0.100 index value. The slowest pace of development is reported in Harishchandrapur-II (0.051) followed by Old Malda, Kaliachak-III, Harishchandrapur-I, Gazole, and others in the district (Fig. 3.3).
2. With 0.196 index value, Chanchal-II block locates the maximum pace of gain under small land size group. Habibpur (0.190), Harishchandrapur-I (0.160), Manikchak (0.152), Ratua-I (0.145), and Kaliachak-I, Kaliachak-II, and Ratua-II have also added more than 0.100 index in the district. Harishchandrapur-II (0.037) adds least, while Kaliachak-III remain adds under 0.050 index (Fig. 3.5).
3. The blocks like Habibpur (0.219), Chanchal-II (0.176), Harishchandrapur-I (0.152), and others have gained exceptionally well index in semi-medium category. The lowest pace of agricultural development is found in Manikchak (0.026), Old Malda (0.045), Ratua-II, and Kaliachak-I in ascending order (Fig. 3.6).
4. In medium land category, Old Malda (0.243), and Habibpur (0.218) add maximum, while Chanchal-I (0.034) and Gazole (0.044) gain low index. In large category, Bamangola (0.198) and Kaliachak-III (0.068) add medium and low index, respectively.
5. In all land size class, the maximum and minimum pace of development is reported in Chanchal-II (0.168) and Kaliachak-III (0.048) accordingly. The blocks like Habibpur (0.148), Harishchandrapur-I (0.154), Ratua-II (0.154), and others have recorded better pace of development in the district (Fig. 3.9).

The overall agricultural development across land size classes were persistence in Habibpur, Ratua-II, and Chanchal-II blocks in the district. Kaliachak-I, Ratua-I, and Kaliachak-II witness better pace of development under marginal and small land size class. The higher pace in agricultural development is attributed by higher productivity, while crop diversification adds significant index under marginal and small land size categories. The crop diversification-induced development is noted in each land size group; however pace is higher in marginal land size class in the district.

References

Asfaw, A., & Admassie, A. (2004). The role of education on the adoption of chemical fertiliser under different socioeconomic environments in Ethiopia. *Agricultural Economics, 30*(3), 215–228.

Alam, S. (1974). *Planning Atlas of Andhra Pradesh.* Pilot Map Production Plant, Survey of India, Hyderabad, pp.3.

Awulachew, S. B., Merrey, D., Van Koopen, B., & Kamara, A. (2010, March). Roles, constraints and opportunities of small-scale irrigation and water harvesting in Ethiopian agricultural development: Assessment of existing situation. In *ILRI workshop* (pp. 14–16).

Barber, S. A. (1995). *Soil nutrient bioavailability: a mechanistic approach.* John Wiley & Sons.

Batjes, N. H. (1997). A world dataset of derived soil properties by FAO–UNESCO soil unit for global modelling. *Soil Use and Management, 13*(1), 9–16.

Bhadrapur, V. S., & Naregal, S. S. (1992). Levels of agricultural development and its correlates. *New Dimensions in Agricultural Geography: Dynamics of agricultural development, 7,* 197.

Bhat, M. M. (2013). Agricultural land-use pattern in Pulwama district of Kashmir Valley. *International Journal of Economics, Business and Finance, 1*(5), 80–93.

Bhuiyan, C., & Kogan, F. N. (2010). Monsoon variation and vegetative drought patterns in the Luni Basin in the rain-shadow zone. *International Journal of Remote Sensing, 31*(12), 3223–3242.

Devi, R. U., & Govt, S. R. K. (2012). The role of credit co-operatives in the agricultural development of Andhra Pradesh, India. *International Journal of Cooperative Studies, 1*(2), 55–64.

Gibbs, J., & Martin, W. (1962). "Index of Diversification," Taken from Quantitative Techniques in Geography: An Introduction. R. Hammond and P. S. McGullagh, Clarendon Press, Oxford. P. 21.

Ghose, A. K., & Saith, A. (1976). Indebtedness, tenancy and the adoption of new technology in semi-feudal agriculture. *World Development, 4*(4), 305–319.

Husain, M. (1996). *Systematic agricultural geography.* Jaipur: Rawat Publications.

Jathar, G. B., & Beri, S. G. (1949). *Indian economics; A comprehensive and critical survey.* London: Oxford University Press.

Johnston, B. F., & Mellor, J. W. (1961). The role of agriculture in economic development. *The American Economic Review, 51*(4), 566–593.

Joshi, Y. G., & Dube, J. (1979). Measurement of regional disparity in the level of agricultural development in Madhya Pradesh. *The Deccan Geographer, India, XVII*(3), 585–593.

Körner, O., & Challa, H. (2003). Process-based humidity control regime for greenhouse crops. *Computers and Electronics in Agriculture, 39*(3), 173–192.

Kumar, N. P. (2015). Disparities in agricultural development of Uttar Pradesh: An inter district study. *Indian Journal of Regional Science, 47*(2), 141–156.

Mathur, A. S., Das, S., & Sircar, S. (2006). Status of agriculture in India: trends and prospects. *Economic and political weekly,* 5327–5336.

Mundinamani P.S., (1985). Agricultural development in Karnataka (Ph.D. Theses), Karnataka University, Dharwad. pp.324.

Nath, V. (1969). The growth of Indian agriculture: A regional analysis. *Geographical Review of India., IX,* (3), 369. 14.

OECD. (1999). *The agricultural outlook 1999–2004,* Paris, France.

Pretty, J., & Ward, H. (2001). Social capital and the environment. *World Development, 29*(2), 209–227.

Ram, P. (1989). An analysis of correlation matrix in agricultural development: A case study of Mirazapur district. *The Indican Journal of Ceography,* 77–85.

Raza, M. (1980). Regional Developments Analytical framework and Indicator. *Indian Journal of Regional Science.* Vol. VII, No. I, pp.11–34.

Robbins, C. W., Meyer, W. S., Prathapar, S. A., & White, R. J. G. (1991). Understanding salt and sodium in soils, irrigation water and shallow groundwaters. *Water Resources.* Series 4.

Sarkar, A. (2013). *Quantitative Geography: Techniques and Presentations.* Orient BlackSwan.

Sen, A. (1981). Market failure and control of labour power: Towards an explanation of 'structure' and change in Indian agriculture. Part 1. *Cambridge Journal of Economics, 5*(3), 201–228.

Siddiqui, S. H. (2007). *Fifty years of Indian agriculture*. New Delhi: Concept Publishing Company.

Siddiqui, S. H., et al. (2011). Technology and levels of agricultural development in Aligarh District. *The Geographer, 58*(2), 12–16.

Singh, A. K. (1987). *Agricultural development and rural poverty*. South Asia Books.

Singh, J. (1976). An agricultural geography of Haryana. *Vishal Publication*. Kurukshetra. pp. 376–384.

Tewari, R. T., & Singh, N. (1985). Development and productivity in Indian agriculture: A cross-section temporal analysis. *Indian Journal of Regional Science, 17*(1), 65–75.

Theurillat, J. P., & Guisan, A. (2001). Potential impact of climate change on vegetation in the European Alps: A review. *Climatic Change, 50*(1-2), 77–109.

Yang, W. (1965). Methods of farm management investigations for improving farm productivity (No. 04; RMD, S401. U6 Y3.).

Chapter 4
Status of Crop Diversification

In the last chapter, we have found that non-physical inputs and crop diversification are positive and strongly associated. And, the degree of association becomes stronger and more robust from 1995–1996 to 2015–2016. It means if non-physical inputs further increase, then crop diversification will rise. Beyond inputs-outputs relation, the present chapter is framed to know the trends and current status of crop diversification crop category-wise under different land size groups. The analysis is presented in the following sequence. The theoretical aspects of crop diversification including factors, hindrances and techniques have presented national and international importance. This follows the trends, patterns and the extent of crop diversification at the block level in the district. And next to this, the extent and change in crop diversification under different crop categories across land size classes is displayed and analysed. The last section examines the pattern and reasons of crop diversification through field data.

4.1 Concept of Crop Diversification

Agricultural diversification is one of dimensions to check whether agricultural development of a region between two points of time has happened or not. The word 'diversification' is derived from a Greek word 'diverge' which means change from single point to another direction. In this sense, diversification declines the dominance of one instead of enhancing the shares of other crops or activities. Hence, agricultural diversification is thereby stated as decrease of intensification and allows other activities or shifts from one enterprise to another enterprise in an agricultural system. In other words, agricultural diversification means growing crops, livestock farming, bee keeping, fishing and others which increases the share of high-value activities in existing agricultural system. No doubt, the agricultural diversifications have many goods over intense specialized farming. Output measure of diversification suggests that the higher-income option, employment generation, risk reduction, food security, sustainable agriculture, optimization of resource use, higher net returned, high land use efficiency and secure farming are many positives unlike traditional subsistence farming. On the other hand, crop diversification is one of the types of agricultural diversification enjoying identical vantages over subsistence agricultural diversification. The recent experiences from Asia (particularly South Asia and Southeast Asia), Middle East and North Africa show an increasing focus on crop diversification through policy intervention to attain the objectives of output growth, employment, income and natural resources stability (Siddiqui and Rahaman 2016).

Crop diversification as a type of agricultural diversification is more suitable and easier to adopt because it requires minimum resources investment in terms of money, place, workforces, time and technologies. Therefore, in a country like India, where marginal and small farmers are more in number but do not have dedicated market infrastructure and technologies; specially storage, dedicated transport, improved seeds, agro-processing industries, crop diversification ensures more ease than agricultural diversification.

In agriculture geography, crop diversification simply means growing multiple crops in a year on particular field or it is an agricultural system where farmers harvest varieties of crop instead of one. A plethora of terms associates with it at different cropping levels such as mixed farming which refer to growing multiple crops subsequently at a time in the same field (Shucksmith and Smith 1991), shifting away from monoculture (Newby 1983), "pluriactivity" (Fuller 1990), "crop varieties" (Petit et al. 1993) and "crop substitution and crop adjustment" (Gunasena 2001). In agricultural cropping system, "crop diversification firstly denotes to species diversification of cultivated crops and, secondly refers to the diversification of varieties and ecotypes of same varieties to maximise income and value-added processed product" (Mengxiao 2001).

The cultivators from developing countries, for example, Chinese farmers, adopted this practice long ago as population pressure compelled to be self-reliant with physical resource constraint. The traditional practice of to be self-reliant through crop diversification in Malaysia was focused on both horizontal diversification and vertical diversification. The horizontal diversification means increase in number of crops in exiting cropping system while upstream and downstream activities which are related to add the value to primary crop are termed as vertical diversification (Yahya 2001). The vertical variant emphasizes on inter-sector linkage and developed competitive value chain in production system. The value production system in developed countries has already witnessed a strong linkage with crop diversification. The contemporary advance farming system in Europe, Americas and Australia is based on crop diversification measures especially found in urban truck farming, horticultural gardening, market farming and sites farming which ultimately links value production systems.

4.2 Drivers of Crop Diversification

The diversity-induced diversification has gained impetus from natural as well as human environments. The ecological difference in natural environment along with climate and consumption habit creates a possibility of crop diversification (Sharma 2001). In a given situation, a variety of agricultural products are commercially feasible and locally suitable which is mostly driven from natural environment. The different geo-climatic condition guides commercial entrepreneurship in agriculture because different climatic conditions provide opportunity to grow diversity of crops throughout the year. Therefore, crops grown in hilly tract would be profitable if the same are sold in other regions in stipulated time. In this context, crop diversification is considered as instrument to grow best profitable commodity to earn money from it. And, then the judicious use of land makes agriculture commercially viable to farmers which is also nature driven.

The non-physical milieu such as growing consumerism at hotel and restaurants, export opportunities, new technology, increasing population and government policy have also shaped the outcome and context of crop diversification. The economic context is more focused among non-physical opportunities of diversification than others. In India, several attempts have been made to increase farmer income through crop diversification measures where emphasis has been given in shift of traditionally less remunerative crops to more remunerative crops through policy initiative (Hazra 2001). In a holistic approach of crop diversification including policy, export opportunities, market and others defined "crop diversification as a strategy of shifting from less profitable to more profitable crops, in changing variety of cropping systems, increasing export in domestic and international market, protecting environment and making favourable conditions for agriculture- fishery-forestry-livestock" (Van Luat 2001).

In Chap. 3, we have found agricultural development positively associated with crop diversification. Therefore, variables and indicators of agricultural development along with market infrastructure and government policy are important drivers in expansion of crop diversification in a region. Pitipunya (1995) in his study entitled "Determinant of Crop Diversification on Paddy Field: A Case Study of Diversification to Vegetables" has tried to find out the determinants of diversification from paddy to vegetables. The study is based on primary data which is collected from the household's survey. The study used a logit model for analysing the impact and importance of infrastructural facilities (family labour, education, trading experience, level of information, man-land ratio, depreciation) to promote diversification. Pingali and Rosegrant (1995) have discussed that the commercialization and diversification of farming have uprooted the scope of the integrated farming system. In commercial farming, entrepreneurs are specialized in diversified mode due to the growing space of urban area and economic growth. The agricultural labour forces hence shift towards the growing space of urbanization. Appropriate government policies and efficient rural infrastructure (crop improvement research and extension and water management) can reduced the possible adverse consequence of commercialized farming and increase diversification in rural areas.

Girish and Mehta (2003) have examined the relationships between crop diversification and socio-economic factors (farm size, land tenure, education, distance from town/market and other important factors) by use of primary and secondary sources of data. The result shows that the higher level of agricultural diversification lies in developed villages and lower in backward (completely non-irrigated) villages. The village having opportunities of economic (farm size, extent of tenancy, bullocks' cart, tractors, sources of farm and non-farm income) and social (like family size, age, education of farmer, and distance from market) accessibility reports better diversification. The Herfindahl and entropy indices show that the diversification extends in better economic accessible villages, while regression summary suggests that economic variables are determinant in explaining the diversification indices. A study has done by Shafi (2008) on "diversified cropping pattern and agricultural development of Dadri block, U.P" based on secondary data. The factor matrix is used to show agricultural development and crop diversification in study area. The changes in cropping pattern along with techno-institutional modernization are highly suggested for better diversification and development. Chakraborty (2013) has analysed the relationship between intensity of irrigation and crop diversification of Murshidabad District on two points of times (1996–1997 and 2006–2007) based on secondary data. To measure crop diversification and irrigation intensity, he has used Jasbir Singh (1976) index of crop diversification and Ram's technique (1979) of irrigation intensity, respectively. The study revealed that the increase intensity of irrigation leans towards diversification of farming.

Kumar and Gupta (2015) have studied "Crop Diversification towards High-value Crops in India: A State Level Empirical Analysis". The study examined trends in crop diversification towards high-value crops along with identification of major factors determining crop diversification from 1990–1991 to 2011–2012. The high Simpson Diversification Index can be correlated with better cropping intensity, good annual rainfall and increase in gross irrigated area in different years. Mango et al. (2018) have found that the nongovernmental organizations (NGOs) and research organizations have raised crop diversifications directly and indirectly in Malawi of south-east Africa. A study conducted by Mithiya et al. (2018) concludes that the magnitude of rainfall, crop insurance facilities, market density, area under HYV seeds, rural literacy and relative earning from cereal crops have significantly impact on crop diversification. Lancaster and Torres (2019) note that the external (access to market and land) and internal (farmer beliefs and access to information from extension and network sources) factors affect crop diversification of vegetables and fruits growers.

4.3 Constraints of Crop Diversification

To feed growing population, there is negligible scope to expand cultivable area. Under this circumstance, available option will be either crop intensification or diversification or both simultaneously in well-equipped agricultural setup. However, crop diversification has opened up new avenue to farmers, but there are a number of constraints faced by farmers at different intensities from time to time. These are:

1. In present climate change era, unpredictable weather phenomena caused delays in seasons, damage crops and increase burden especially on marginal and small land size holders. The cost of physical constraints on poor peasants puts burden in decision-making to whether cultivation of high-value crops would be feasible nightmares or profit earning sources.
2. The major agronomic constraints such as germination of seeds, viral crop diseases and insect attacks evident from the past have also imprinted as hindrance in way for alternative income generation from high-value crops.
3. The crop diversification demands high-input cost. The economic constraints of diversification hence come in way to it. The market prices during harvest season, trader interest, impact of international market, perishable nature of crops, lack of proper storage facilities and falling minimum support price at local market affect low overall profit margin; therefore, farmers are compelled to shift towards easy and traditional cereal-based cropping pattern.
4. The social constraints of diversification including land tenancy, land ownership, cultural aspects and legal cum economic protection of specific crop have made difficult circumstances in way to crop diversification.
5. From preparation of land to irrigation and proper upbringing to final selling of crops have required keen management practice. The management constraints arise when proper knowledge and information about all aspects of cultivation of high-value crops fail to generate expected income.

Agricultural system in many ways rests in the hand of natural vagaries like rainfall, temperature or any uncertainty. The developed agricultural systems are more resilient to cope up physical unpredictable situation in short time span than subsistence agriculture of developing countries. In given limitations, however, crop diversification as a process is much promising for increasing income of small farmers, generating employments, poverty alleviation, withstanding price fluctuations, enhancing competitiveness, diversified food basket, conservation natural resources, mitigation of aberrant weather effect, nutrition and ecological security and less dependence on off- farm inputs. And therefore, there has been a significant increase in areas under non-food crops and especially under some remunerative crops like vegetables and horticulture in given resources endowment and institutional setup of the world. The needs of improved planting materials for horizontal and vertical diversification have recognized through public and private partnerships at national and international level by the effort of FAO.

4.4   Crop Diversification Techniques

After looking into constraints and vantages of crop diversification, author has attempted to measure crop diversification under different land size categories for two pints of time. To measure diversification at cropping level, there are different indices, namely:

1. Bhatia's method
2. Jasbir Singh's method
3. Herfindahl index
4. Transformed Herfindahl index
5. Gibbs-Martin index
6. Simpson's index
7. Ogive index
8. Gini's coefficient
9. Entropy index (EI)
10. Modified entropy index (MEI)
11. Shannon index
12. Berger-Parker index
13. Composite entropy index (CEI)
14. Crop counting method

The concept of diversification is used differently by geographers, economists and other domain experts. Even within the same discipline, conceptual and methodological differences are very obvious. The nature, approach and method of diversification in different academic pursuits are described hereunder:

1. Diversification as a concept primarily has been used economically from the first half of the twentieth century. The specialization and diversification of manufacturing industries within different cities were calculated by Cleann (1930 cited by Vaishampayan et al. 2019), Mclaughlin (1930), Tress (1938), Reinwald (1949) and others. The diversification of industrial employee was also calculated for future recruitment policy of a company. The economic analyses of diversification employ EI, MEI, CEI and Herfindahl index.
2. In geography, use of diversification index is truly noted in agricultural geography. It was Grimes (1929) who studied the advantages and disadvantages of diversification of agricultural diversification in the United States. Later on, many geographers from the United States, Europe, North America and Asia-Pacific region have studied diversification and specialization status in agriculture, livelihood and other aspects. In geographic analyses, the spatial concern has always been there. The difference in regional variation in diversification on selected aspect aims to measure macro diversification rather than micro diversification as did in economics. The Gibbs-Martin index (1962 cited by Mohammad 1981), Bhatia's method (1965), Jasbir Singh's method (1976) and crop counting methods (1978) are popularly used in geographic analysis.

3. In biological science, the measurement of species diversification within an eco-system or biome or any physical environment has been the major concern of biologist, zoologists, plant pathologist and biochemists. The homogeneity, heterogeneity, evenness and richness in diversification have been measured in different sites. The Simpson's index (1949), Ogive index (1930 cited by Vyas 1996) and Gini's coefficient (1921) are popularly used in biological sciences.
4. The technique of computation of diversification in other disciplines is designed to full fill the objective of the study. In sociology and other social science disciplines, the diversity among social practices and their extent execute through mixed methods depending upon site of study and available data.

To choose appropriate technique for diversification measurement, nature of data (discrete or continuous), types of study (empirical or others), level of study (macro or micro), unit of analyses (more than one at a time or more) and site of study are important dimensions.

In any given point of time, diversification relates to the extent of diversity in crop cultivation or the maintenance of a diverse spread of crops in the cultivated area. Most scholars have assessed the extent of crop diversification by using one or more measures, like (i) number of crops or activities, (ii) Simpson's index of diversity (SID), (iii) Herfindahl index of concentration (HI), (iv) ogive index, (v) entropy index and (vi) composite entropy index. All these indices are constructed using data on the proportionate area under different crops cultivated in a particular area or the proportionate value of agricultural activities in the total value of output. In present study, diversification has been measured by three different approaches, namely:

1. Gibbs-Martin index
2. Simpson's index
3. Crop counting method

These three techniques are best fitted according to nature of data and study. For Gibbs-Martin index, continuous number data is essential which gives best result on proportional area data up to 1% area coverage of individual crop. To compute Simpson Diversification Index, discrete number data is required. And, crop counting method simply counts number of crops for a region which provides the basis of maximum and minimum crop concentration and disperse region.

4.5   Trend and Pattern in Crop Diversification

In present section of the study, Gibbs-Martin and Simpson Index has been used because of nature of data. In the following section, Gibbs-Martin technique used to calculated diversification on the areal extent of crops while evenness of diversification calculated through the Simpson's Index.

To have a precise measure of the degree and homogeneity of diversification the Simpson's Index of Diversification (SID) is used. The Simpson's Index of Diversification (SID) is calculated as:

$$SID = 1 - \sum_{i=1}^{n} P_i^2$$

$$P_i = A_i / \sum_{i=1}^{n} A_i$$

where Pi is the proportion of area under the ith crop and Ai is the actual area under ith crop. The value of the Simpson's Index ranges between 0 and 1. While 0 implies complete specialization, the value 1 signifies maximum diversification.

4.5.1   Trends and Pattern of Crop Diversification in Malda District: Spatial Dimension

Figure 4.1 reveals that crop diversification index in the district has been increasing over time. In the last 20 years, there is increased of 0.162 index value suggesting that the district is moving towards diversification of crops. The pace of changes in index value is higher from 2000–2001 to 2005–2006, and it is least in 1995–1996 to 2000–2001. Though there is again increase in index value from 2005–2006 to 2010–2011 and 2010–2011 to 2015–2016, the pace of increase becomes slow as compared to 2000–2001 to 2005–2006 presumably; it is because of the shift in growing profit from non-food crops. In 1995–1996, the maximum number of crops is grown in Manikchak block which reflects in diversification index. Five blocks, namely, Kaliachak I, II and III, Manikchak and Ratua I, are having more than 0.700 index value signifying that the maximum number of crops are grown over the years. In the same year, block like Habibpur which is in the last ladder of diversification index assumes to have grown one to two crops. Bamangola, Gazole and Harishchandrapur II have medium index value from 0.500 to 0.800. Old Malda and Ratua II are having an index value of less than 0.400. In 2000–2001, three blocks, namely, Kaliachak I, II and III, report the maximum diversification index that is more than 0.800 meaning by the maximum number of crops grown over the year. English Bazar, Manikchak, Gazole and Ratua II blocks locate diversification index from 0.500 to 0.800; hence, they are categorized as medium crop diversified blocks. Bamangola, Chanchal I and II, Habibpur, Harishchandrapur I and II and Old Malda blocks display less than 0.500 index value. In the same year, Old Malda shows the least diversification in the district. In 2010–2011, Kaliachak I and II, Manikchak, English Bazar, Chanchal II and Ratua I have found more than 0.700 index value. This year again, Old Malda is at the bottom of the ranking. In 2015–2016, five blocks, namely, Kaliachak I and II, Manikchak and Ratua I and II, report more than 0.800 index value. Chanchal I, English Bazar and Kaliachak III blocks have found

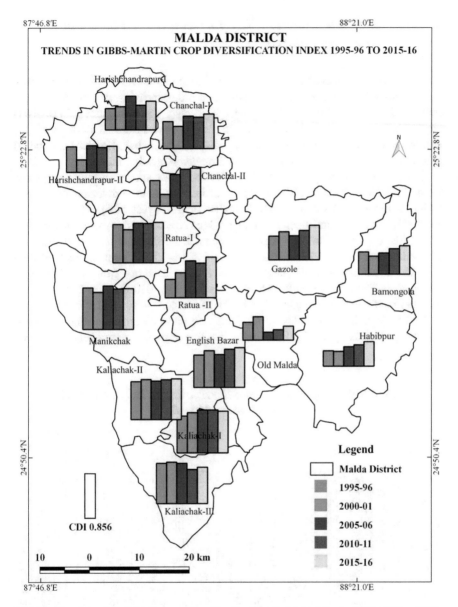

Fig. 4.1 Trends in Gibbs-Martin crop diversification index in Malda District 1995–1996 to 2015–2016. (Source: based on author calculation (data from Agriculture Census))

moderate diversification index between 0.500 and 0.800. In 2005–2006, again Manikchak block is recorded high diversification followed by Kaliachak I and II and Ratua I. Chanchal I and II, English Bazar, Harishchandrapur I and II, Kaliachak II and Ratua I blocks have scored more than 0.500 index meaning thereby moderate diversification.

4.5.2 Trends and Pattern of Crop Diversification in Malda District: By Land Size

The common pattern evolving out from Fig. 4.2 is that the Old Malda and Habibpur blocks consistently posit lower diversification index meaning thereby there are a few crops grown in 1995–1996 and 2015–2016 also. The maximum gain in diversity index is concerned the Chanchal II, English Bazar and Ratua II blocks are in the list. In the overall scenario, three blocks, namely, Manikchak, Kaliachak II and III and Ratua I blocks, are very consistent in crop diversification index.

Land size class category-wise diversification index shows an increasing trend from 1995–1996 to 2015–2016. The pace of increase in crop diversification index under marginal land size category is higher than the rest of the class category. Although small and semi-medium and large land size categories were again successively positive, in medium land size class, there is a decreasing trend in diversification index over 1995–1996 to 2010–2011 and 2015–2016. The maximum increase in index value notes in the large land size class category from 2010–2011 to 2015–2016. The least diversification index is found in 2010–2011 to 2015–2016 in medium land size class category. The land size category-wise diversification proves that crop diversification index across land size class categories in the district is booming which is up to 95% fitted linear line in marginal category (Fig. 4.2).

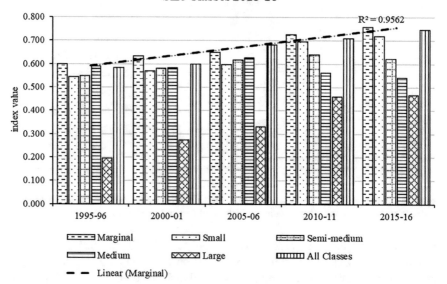

Fig. 4.2 Trends in Gibbs-Martin crop diversification index by land size classes in Malda District 1995–1996 to 2015–16. (Source: Agriculture Census (based on author calculation))

4.6   Homogeneity in Crop Diversification

4.6.1   *Homogeneity in Crop Diversification: Spatial Dimension*

The overall pattern shows that the number of crops has increased between 1995–1996 and 2000–2001 is higher than 2000–2005 and 2000–2010. And, the pace of increase in number of crops is the highest from 2010–2011 to 2015–2016. The total number of crops in 1995–1996 was 25 which reached to 40 in 2015–2016. This increase implies that with the time new crops were introduced by farmers. Although there is an increase in the total number of crops in the district, at block level, there are variations especially in Bamangola, English Bazar, Harishchandrapur I, Kaliachak II and Ratua I and II. In terms of the number of crops grown from 1995–1996 to 2015–2016, it is reported that the Manikchak block gains the maximum number of crops (i.e. 28) from 31 to 59 in the district. Kaliachak II, Habibpur and Harishchandrapur II blocks have gained few crops in the same two decades. These noted blocks also have been consistent in terms of growing number of crops over the years. Other blocks reveal certain level of variation in the trends of crops. Habibpur and Ratua II have been the less consistent block in the district.

Simpson diversification index explain the extent of diversification in terms of variability and homogeneity in the index value. Increase number of crops over the year does not guarantee that the index value will increase proportionally; rather it is much affected by the maximum and the minimum difference in number of crops.

Table 4.1 shows that there is increased in the number of crops in district across different block, but Table 4.2 displays a decreasing trend in index value. It signifies

Table 4.1 Trends in number of crops in Malda district

Blocks	1995	2000	2005	2010	2015
Bamangola	16	15	23	28	30
Chanchal I	25	29	26	42	45
Chanchal II	23	25	13	41	44
English Bazar	28	29	33	39	42
Gazole	24	36	39	29	39
Habibpur	22	19	19	21	27
Harishchandrapur I	24	24	26	35	43
Harishchandrapur II	25	23	27	29	31
Kaliachak I	22	27	23	50	45
Kaliachak II	27	25	23	16	27
Kaliachak III	32	32	42	44	48
Manikchak	31	34	23	59	59
Old Malda	25	21	20	32	39
Ratua I	26	30	33	47	49
Ratua II	27	20	22	15	27
District	25	26	26	35	40

Source: Agriculture Census, Government of India, 1995–1996 to 2015–2016

Table 4.2 Trends in Simpson Crop Diversification Index at block level in Malda district

Blocks	1995	2000	2005	2010	2015
Bamangola	0.939	0.829	0.739	0.844	0.823
Chanchal I	0.905	0.733	0.843	0.823	0.827
Chanchal II	0.900	0.806	0.821	0.815	0.816
English Bazar	0.870	0.830	0.872	0.780	0.798
Gazole	0.890	0.784	0.798	0.837	0.829
Habibpur	0.877	0.762	0.779	0.805	0.800
Harishchandrapur I	0.895	0.843	0.895	0.824	0.838
Harishchandrapur II	0.876	0.862	0.821	0.833	0.830
Kaliachak I	0.876	0.818	0.870	0.820	0.830
Kaliachak II	0.889	0.881	0.860	0.742	0.765
Kaliachak III	0.897	0.881	0.844	0.837	0.838
Manikchak	0.897	0.834	0.877	0.833	0.842
Old Malda	0.905	0.873	0.852	0.726	0.751
Ratua I	0.881	0.851	0.882	0.820	0.832
Ratua II	0.871	0.816	0.826	0.876	0.866
District	0.860	0.821	0.829	0.807	0.812

Source: Computed by author

that there is increasing gap in the maximum and the minimum number of crops grown in different blocks. In 1995–1996 the maximum and minimum number of crops gap was 16, and it became 32 in 2015–2016. In overall, Simpson diversification illustrates there is an increasing gap in diversification among the blocks in the district.

4.6.2 Homogeneity in Crop Diversification: By Land Size Classes

The number of crops in different land size class categories from 1995–1996 to 2015–2016 illustrates an increasing trend in the district. Except the large land size class category, all other categories showing a linear growth in terms of the number of crops grown in the district in said period. In marginal land size class category, the maximum increase has been recorded which is most consistent in trends also. The small land category and semi-medium land size group have found a certain level of variations although there is a gain in 2015–2016 (Table 4.3). The all land size class category in the district locates an addition of 23 crops from 1995–1996 to 2015–2016.

In case of land size class category (Table 4.4), Simpson diversification index again depicts decreasing trend. The maximum decrease is noticed in semi-medium land size class followed by medium, small, large and all classes categories. The maximum increases in number of crops report under marginal land size class in both

Table 4.3 Trends in number of crops by land size category in Malda district

Land size class	1995–1996	2000–2001	2005–2006	2010–2011	2015–2016
Marginal	31	64	68	65	83
Small	27	48	44	37	54
Semi-medium	24	40	46	30	48
Medium	23	25	35	17	39
Large	8	4	0	5	6
All classes	33	69	74	66	84

Source: Agriculture Census, Government of India, 1995–1996 to 2015–2016

Table 4.4 Trends in Simpson Crop Diversification Index by land size class category in Malda district

Land size class	1995–1996	2000–2001	2005–2006	2010–2011	2015–2016
Marginal	0.892	0.843	0.856	0.828	0.859
Small	0.903	0.829	0.851	0.772	0.857
Semi-medium	0.909	0.794	0.860	0.791	0.871
Medium	0.913	0.830	0.876	0.794	0.860
Large	0.857	0.833	0.000	0.900	0.800
All classes	0.890	0.838	0.856	0.830	0.859

Source: Computed by author

the years. Although the number of crops rises, index value decrease, and it is because of increasing gap in number of crops between 1995–1995 and 2015–2016 that caused a decrease in index value.

Gibbs-Martin crop diversification index observes increasing trends in index value over the years at both block level and land size class category level. The nature of change in index value is not homogenous across blocks rather some block experiences huge transformations. The land size category (Table 4.4) index also suggests an ascendency in development in index value under marginal land size class category. However, small and semi-medium category is found significant improvement, but their index value is lower than the marginal land size class. So, diversification in the district is dominated under marginal and small land size category.

In terms of number of crops increasing over the year is on the positive side at block level as well as land size class category. There are four blocks that are very constant concerning to growing addition in the total number of crops at block level. The overall district scenario shows increase of almost double number of crops from 1995–1996 to 2015–2016. The marginal land size class category reports the maximum number of crops which offer sufficient information with regard to the marginalization of improvement in the number of crops in district.

Gibbs-Martin diversification tends to discover a booming picture of crop diversification across land size categories at block level. Simpson index reveals an increasing disparity in trends of crop diversification across land size categories at block level in the district.

4.7   Crop Diversification Under Different Crop Categories

This section tries to explain crop diversification under different crop categories such as cereals, pulses, vegetables, spices, fruits and others across land size classes at block level.

4.7.1   Crop Diversification Under All Crops

Figure 4.3 reveals that except medium land size class category, all other land size classes in the district display increase in diversification index under all crops from 1995–1996 to 2015–2016. And, it is mainly due to improved non-physical infrastructure. The maximum gain is noted in Ratua II block under marginal, small and semi-medium land size classes with index value 0.307, 0.307 and 0.445, respectively. Kaliachak III (0. 401) and Bamangola (0.326) blocks have gained significant index value under medium and large land size classes. In marginal category, the blocks like Harishchandrapur II, Kaliachak II, Old Malda and Manikchak posit negative change in index value. The maximum decrease is found in medium land size category under Manikchak (0.522) block.

4.7.2   Crop Diversification Under Cereals

The change in crop diversification under cereal crops in the district shows that there is very minimum increase over years across different land size classes except the medium land size category (Fig. 4.4). Other than Bamangola, Chanchal II, Gazole, Habibpur, Kaliachak I, Ratua I and Ratua II, the other remaining blocks locate the negative change in marginal land size category in two decades. The small, semi-medium and all land size groups also report negative change other than those said block in marginal land size category. The maximum gain in cereal crops locates under medium land size group in Gazole (0.339) block, and the maximum negative change displays in Kaliachak II (0.518) under medium land size class. The profit from cereal crops is comparatively low than other crops such as oilseeds, pulses, vegetable and others; therefore, farmers tend to decrease area under this crop category. The frequent attacks from insects and pests, flood and drought-like situation are also responsible in decreasing diversification under this crop category.

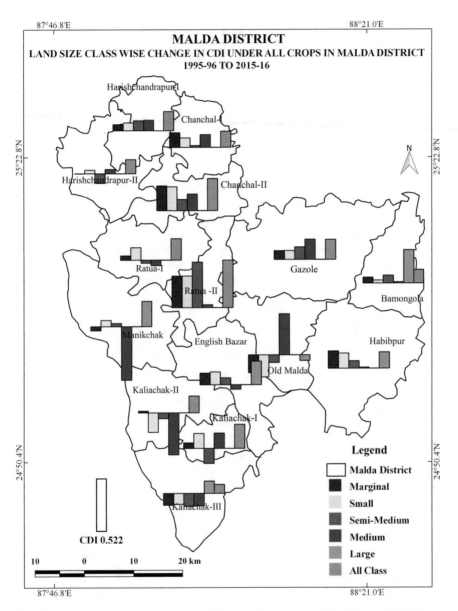

Fig. 4.3 Land size class wise change in CDI under all crops in Malda District 1995–1996 to 2015–2016. (Source: See Annexure 4a)

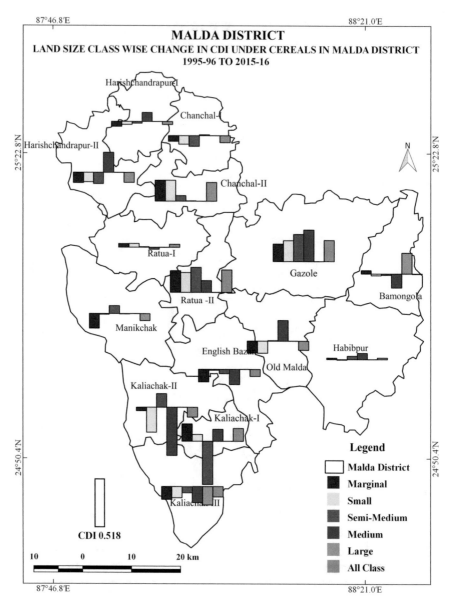

Fig. 4.4 Land size class wise change in CDI under cereals in Malda District 1995–1996 to 2015–2016. (Source: see Annexure 4b)

4.7.3 Crop Diversification Under Pulses

The change in diversification index under pulses reveals that the district has experienced negative change under marginal, small and all land size classes from 1995–1996 to 2015–2016. Although there are improvements in index value under semi-medium category, it is very low, i.e. 0.074 (Fig. 4.5). The maximum gain is noted in Chanchal I (0.574) in small class followed by Bamangola and Chanchal I block under marginal and semi-medium land size categories, respectively. The maximum decrease is found in Gazole (0.710) block under all land size classes category. Some of the block reports mix pattern in different land size classes such as in Kaliachak II, Ratua I and Kaliachak I in the district. The low productivity, prevalence of insects and pest and recurrence of diseases are important concern related with low gross crop area and diversification among pulse crops in the district (CDAP XII 2015).

4.7.4 Crop Diversification Under Spices

There is increase in diversification index under spices in all land size class groups in the district (Fig. 4.6). Among major spices crops, chilly and onion stake more than half of the total GCA. Due to lack of storage facility for onion crop, farmers recently have been shifting towards other spice crop; hence diversification is increasing in district. At the block level, the maximum gain reports under small land size class in Chanchal I (0.668) followed by Harishchandrapur I (0.622), Kaliachak I (0.531), English Bazar (0.142) and Kaliachak II (0.598) blocks in marginal, small and all land size classes correspondingly. The maximum negative growth is observed under small land category in Ratua II, Kaliachak II (0.174) and Ratua II (0.195) in ascending order. Although spice crop offers high monetary return, it is labour-intensive and weather critical crops. There is often risk of crop damage due to natural calamities.

4.7.5 Crop Diversification Under Fruits

The extent of diversification under fruits in marginal and all land size class groups in district shows the negative growth over 20 years. The lack of agro-based specifically fruit-based industries hinders expanding diversification in the district. In marginal land size class, Harishchandrapur II block adds the maximum index (0.793) and remained the most diversified block in the district. Kaliachak II (0.424), Harishchandrapur I (0.460) and Kaliachak I (0.334) display positive change in index to 2015–2016 from 1995–1996 (Fig. 4.7). Recent change in share of GCA under mango and litchi in the Tal region of district is decreasing due to prices instability. The maximum decline in index value is seen under semi-medium class in

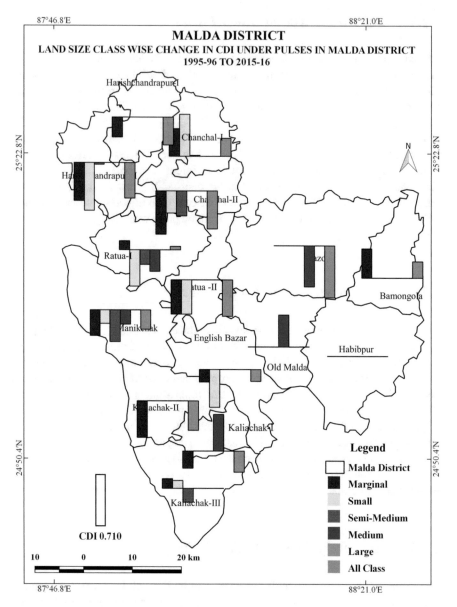

Fig. 4.5 Land size class wise change in CDI under pulses in Malda District 1995–1996 to 2015–2016. (Source: see Annexure 4c)

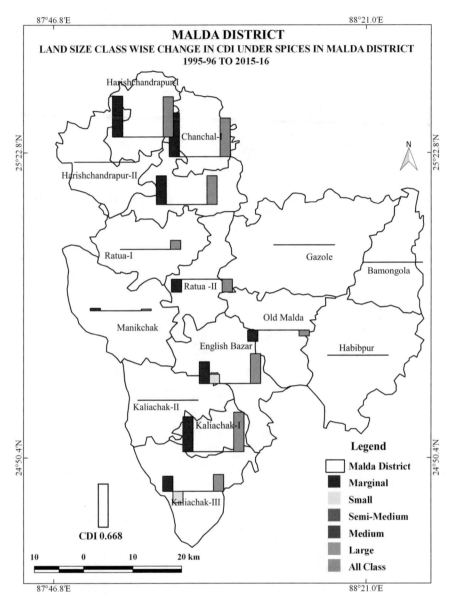

Fig. 4.6 Land size class wise change in CDI under spices in Malda District 1995–1996 to 2015–2016. (Source: see Annexure 4d)

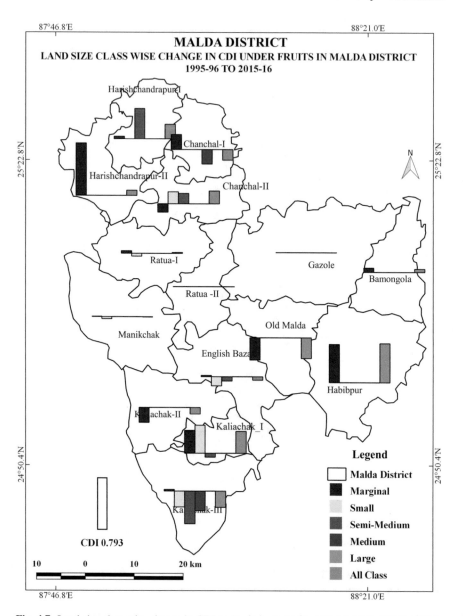

Fig. 4.7 Land size class wise change in CDI under fruits in Malda District 1995–1996 to 2015–2016. (Source: see Annexure 4e)

Kaliachak III. English Bazar, Chanchal I and II, Kaliachak I and II, Manikchak and Ratua I show mix pattern under different land classes. The mango and litchi have been main fruits in the district. The diversification under fruits crop is expected to be increased only after possible vertical diversification is created in the district.

4.7.6   Crop Diversification Under Vegetables

Though the GCA under vegetables crops is decreased substantially in the district, still this crop category represents as the most diversified group because it requires minimum area to cultivate. The maximum gain in diversification index under vegetable crops is located in the district under marginal class (0.430) followed by small (0.123), medium (0.297) and all land size classes (0.394). The negative change appears in semi-medium class (0.235) in the district (Fig. 4.8). In marginal land size class, the maximum increase is recorded for Kaliachak I (0.855) followed by Harishchandrapur II (0.626), Kaliachak III (0.677), Manikchak (0.571) and English Bazar (0.587). The highest gain is reported under medium (0.721), semi-medium (0.408), large (0.500) and all land size class (0.903) in Kaliachak I, Kaliachak II, II and III, respectively.

4.7.7   Crop Diversification Under Oilseeds

More than 70% of total GCA is occupied by mustard and rapeseeds in district. The decreasing trend in diversification under oilseed crop is due to small land size and land tenancy in Malda district. Figure 4.9 shows that diversification index under oilseeds is decreased across different land class size categories in the district. The maximum decrease is displayed under small land size class followed by semi-medium, medium, all land size and marginal class. In the marginal category, Chanchal I block depicts the maximum decrease in diversification index that is 0.404. Old Malda block in small category and Chanchal I in semi-medium, medium and all size class have recorded the maximum decrease. Only one block in the district, i.e. Chanchal II represents positive growth in index value from 1995–1996 to 2015–2016 (Fig. 4.9).

4.7.8   Crop Diversification Under Fibre

Figure 4.10 represents evidence on crop diversification index under fibre. Among fibre crops, jute shares more than 80% GCA followed by mesta and cotton in the district. In district, the diversification under said crop category is decreased from 1995–1996 to 2015–2016. Overall diversification index seems that the district becomes one crop dominated fibre crops, except Chanchal I, Habibpur, English Bazar and Ratua I block. The maximum decrease in diversification index is found in Kaliachak III block under marginal (0.454), small (0.304), semi-medium (0.293), medium (0.474) and all land size classes (0.399). The maximum increase under small land size class is recorded in Chanchal I (0.072) block. The positive increase in index value also appears in marginal (0.038), semi-medium (0.037) and all land size (0.023) classes in Kaliachak I, Habibpur and Kaliachak I block, respectively.

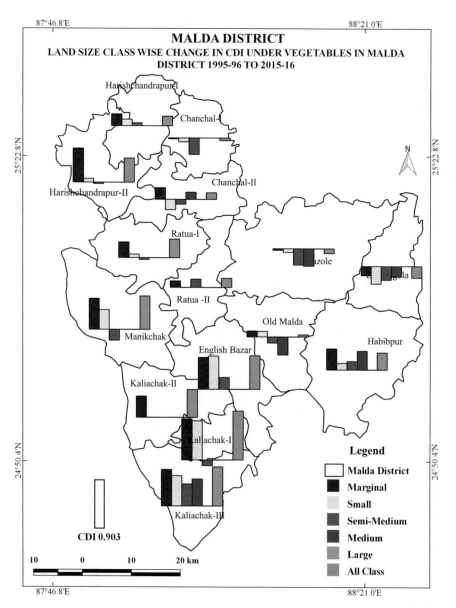

Fig. 4.8 Land size class wise change in CDI under vegetables in Malda District 1995–1996 to 2015–2016. (Source: see Annexure 4f)

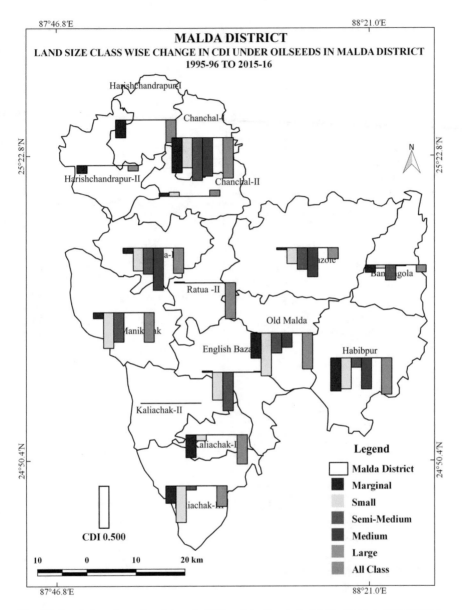

Fig. 4.9 Land size class wise change in CDI under oilseeds in Malda District, 1995–1996 to 2015–2016. (Source: see Annexure 4g)

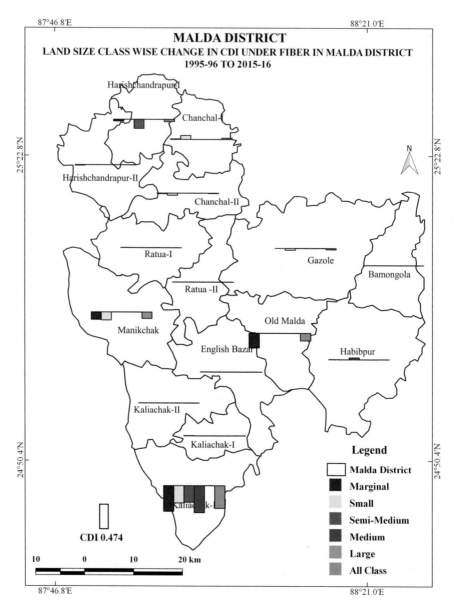

Fig. 4.10 Land size class wise change in CDI under fibers in Malda District 1995–1996 to 2015–2016. (Source: see Annexure 4h)

4.8   Crop Diversification: Crop Count Method

One of the problems with Gibbs-Martin diversification index is that if the area occupies by an individual crop is less than 1% in total GCA in the block or district, then share of a particular crop does not reflect in diversification index. In such case, real representation of crop diversification is shaded by dominate crop, and therefore, it is necessary to count the number of crops grown at spatial-temporal context. If the number of crops under each crop category increase or decrease, then it signifies the real situation of diversification in selected space over time. An attempt has been made to count and present the number of crops under each crop's categories between two points of time across land size category at block level in the district. Figure 4.11 represents the total number of crops under each crop's category grown in 1995–1996 and 2015–2016.

Comparative analysis reveals that the total crops under all crops categories was 34 in 1995–1996 which raised to 57 in 2015–2016. In the same period, the number of crops under different crop categories has increased like in cereals up to seven, pulses up to six, spices up to six, fruits up to nine, vegetables up to 20, oilseeds up to six and fibre up to three from six, three, three, four, nine, five and four crops individually in the district. In 1995–1996, the maximum number of cereals crops records in English Bazar, i.e. six. Except Bamangola, Kaliachak I and Kaliachak II block, all other blocks produced three crops which was the maximum in pulses crop category. In the same year, the maximum number of crops was recorded in Manikchak (03) in spices, Manikchak (04) in fruits and Chanchal II and Harishchandrapur II (04) in fibre crops in district. With nine and five crops under vegetables and oilseeds, the district has recorded the maximum under different blocks. In 2015–2016, the blocks like Gazole (07), Manikchak (06), Chanchal I (06), Manikchak and Ratua I (09) and Manikchak (20) show the maximum number of crops under cereals, pulses, spices, fruits and vegetables, respectively. Manikchak (06) and Chanchal I (03) blocks locate the maximum increase under oilseeds and fibre crops correspondingly.

4.8.1   Change in Number of Crops 1995–1996 to 2015–2016: Spatial Dimension

Figure 4.11 reveals that the maximum number of crops has increased under vegetables followed by fruits and spices in the district from 1995–1996 to 2015–2016. The maximum number of crops, i.e. 12, is increased in Kaliachak I, while Manikchak (11), Kaliachak III (10) and Old Malda, Chanchal I, Chanchal II, Ratua I and Ratua II block (eight each) have also added positively. In cereal crops, Bamangola (03) shows the maximum increase followed by Gazole (02), Manikchak (02) and Chanchal I, Chanchal II, Harishchandrapur I and Kaliachak I block in the district. Under pulses, Manikchak and Kaliachak I blocks have displayed increase of three

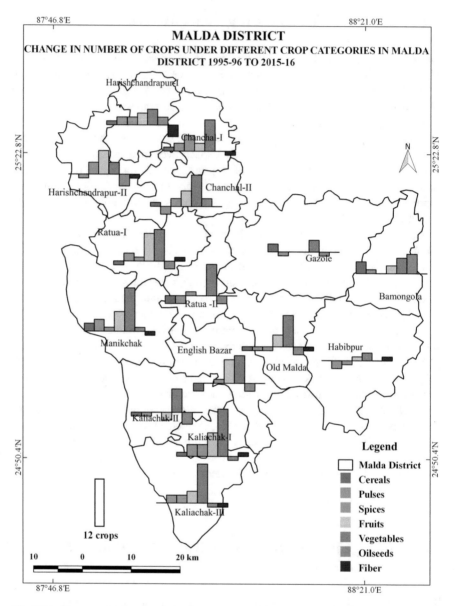

Fig. 4.11 Change in number of crops under different crops categories in Malda District 1995–1996 to 2015–2016. (Source: see Annexure 4i)

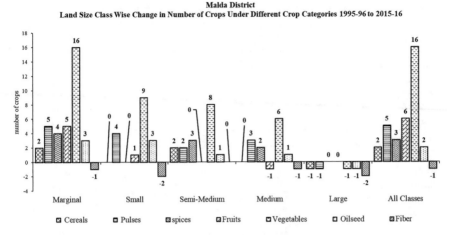

Fig. 4.12 Land size class wise change in number of crops under different crop categories in Malda District 1995–1996 to 2015–2016. (Source: based on Table 4.5)

crops, while in spice it is Chanchal I that gain the maximum, i.e. four crops. Harishchandrapur II and Kaliachak I block add three each more crops under spice category. Ratua II gains the maximum increase of seven, followed by Harishchandrapur II and Kaliachak I (six each), while Manikchak and Chanchal II enhance more five and four crops, respectively. Bamangola block adds five crops under oilseeds. Chanchal II and Harishchandrapur I increase two crops each. In fibre crops category, the maximum gain is just one crop in Habibpur, Kaliachak I and Ratua I block in the district. The maximum decrease is found under oilseeds, fibre and pulse crops from three to four. In oilseeds category, Harishchandrapur I and Kaliachak II have loosed three crops each in the district. Harishchandrapur I lost three crops which is maximum under fibre crops category.

4.8.2   Change in Number of Crops 1995–1996 to 2015–2016: By Land Size Classes

Land size class category-wise data on the number of crops from 1995–1996 and 2015–2016 reports that except fibre, the number of crops under each crop categories has increased in the district (Fig. 4.12). The maximum crop increase is noticed under vegetable crops, i.e. 16, followed by fruits 6, pulses 5, spice 3, vegetable and cereals (two each). Except fibre crops, marginal, small and semi-medium land size classes have added crops under each crop categories in two decades. Medium and large land size classes display mix pattern, in which vegetables, pulses and spices add only in medium land class, but large category shows negative growth in district.

Table 4.5 Number of crops under different crops categories in Malda district

Size class	Cereals	Pulses	Spices	Fruits	Vegetables	Oilseeds	Fibres	Total crops
1995–1996								
Marginal	6	3	3	4	9	5	5	35
Small	5	3	3	4	8	3	4	30
Semi-medium	4	3	2	3	6	4	3	25
Medium	4	3	2	3	6	3	3	24
Large	3	1	0	0	3	1	2	10
All classes	6	3	4	4	9	6	5	37
2015–2016								
Marginal	8	8	7	9	25	8	4	65
Small	5	7	3	5	17	6	2	43
Semi-medium	6	5	5	3	14	5	3	38
Medium	4	6	4	2	12	4	2	32
Large	2	0	0	0	2	0	0	4
All classes	8	8	7	10	25	8	4	66

Source: Agriculture Census, Government of India

4.9   An Insight from the Field

4.9.1   Farmer Who Diversify Crop

Out of total respondent from the district, 86% farmers respond that they do diversify their crops very often (within less than 3 years). The nature of diversification is mostly from food to non-food cash crop which promises higher income. Primary survey also reports that 78.70% of farmers have diversified their crops dominated under vegetables, flowers and other non-food crops. Crop diversification is low under medium land size class category that is 66.67% due to leased-out land and low profit due to labour and technological cost. The semi-medium land size farmer (80%) reports diversification under pulses, oilseeds and vegetables because of mainly profit opportunities. The farmers having land size more than 1 ha prefer monoculture because high cost of production. The marginal and small land size farmers do depend on diversification mostly from vegetables, oilseeds, pulses, floriculture and fruits. The intensive cultivation backed by fertilizer and HYV seeds hike crop diversification under marginal (87.50%) and small (84.44%) land size farmers (Fig. 4.13).

4.9.2   Reasons of Crop Diversification

The reason of crop diversification has been presented in Fig. 4.14. It is observed that profit as a reason contributes around 49.10% in the total causes of district. The family consumption adds around 17.83% followed productivity (17.05%), risk

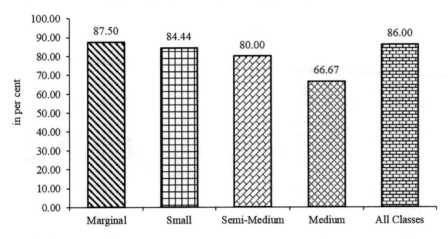

Fig. 4.13 Percentage of farmer who diversify crops very often in Malda District 2018. (Source: based field survey 2018)

aversion (9.82%) and climate change (6.20%). The efficient and available labour (contributed by male and female) in vegetables, spices and fruits cultivation do encourage high profit opportunity in case of marginal and small land size holders. The risk from natural calamities such as flood, crop disease, monsoon rainfall variability and hailstorm has alternated diversification most in medium land size class category.

The changing season of rainfall do change cropping pattern and crop diversification. The climate change and climate variability do impact negatively in crop diversification. The higher productivity of specific crop also attracts the farmer in decision-making for diversification which is maximum under medium land size category. The district is exposed of different natural calamities such flood, drought, storm and hailstorm, rainfall variability and crop diseases. And therefore, farmer practices intensive cultivation to minimize risk in farming.

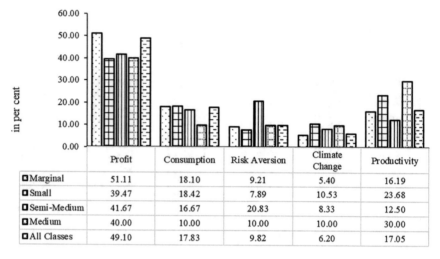

Fig. 4.14 Reasons for crop diversification in Malda District 2018. (Source: based field survey 2018)

References

Bhatia, S. S. (1965). Patterns of crop concentration and diversification in India. *Economic Geography, 41*(1), 39–56.

Chakraborty, A. (2013). Relating irrigation to crop diversification: A spatial-temporal study in Murshidabad District, West Bengal (1996–97 to 2006–07). *Indian Streams Research Journal, 2*, 12.

Deputy Directorate of Agriculture, Malda. (2015, March). *Comprehensive District Agricultural Plan (C-DAP)*. NABARD Consultancy Services Pvt Ltd.

Fuller, A. M. (1990). From part-time farming to pluriactivity: A decade of change in rural Europe. *Journal of Rural Studies, 6*(4), 361–373.

Gibbs, J., & Martin, W. (1962). *Index of diversification* (p. 21). Taken from quantitative techniques in geography: An introduction. R. Hammond and PS McGullagh. Oxford: Clarendon Press.

Gini, C. (1921). Measurement of inequality of incomes. *The Economic Journal, 31*(121), 124–126.

Girish, M., & Mehta, P. (2003). Crop diversification: An empirical analysis on Kangra farms of Himachal Pradesh, India. *Asia-Pacific Journal of Rural Development, 13*(2), 1–18.

Grimes, W. E. (1929). Diversification of agriculture—Its limitations and its advantages. *The Annals of the American Academy of Political and Social Science, 142*(1), 216–221.

Gunasena, H. P. M. (2001). Intensification of crop diversification in the Asia-Pacific region. *Crop Diversification in Asia Pacific Region*.

Hazra, C. R. (2001). Crop diversification in India. In M. K. Papadimitriou & F. J. Dent (Eds.), *Crop diversification in the Asia-Pacific Region* (pp. 32–50). Bangkok: Food and Agriculture Organization of the United Nations. Regional Office for Asia and the Pacific.

Kumar, S., & Gupta, S. (2015). Crop diversification towards high-value crops in India: A state level empirical analysis. *Agricultural Economics Research Review, 28*(2), 339–350.

Lancaster, N. A., & Torres, A. P. (2019). Investigating the drivers of farm diversification among US fruit and vegetable operations. *Sustainability, 11*(12), 3380.

Mango, N., Makate, C., Mapemba, L., & Sopo, M. (2018). The role of crop diversification in improving household food security in Central Malawi. *Agriculture & Food Security, 7*(1), 7.

McLaughlin, G. E. (1930). Industrial diversification in American cities. *The Quarterly Journal of Economics, 45*(1), 131–149.

Mengxiao, Z. (2001). *Crop diversification in China* (pp. 24–31). Bangkok: Crop Diversification in the Asia-Pacific Region. FAO Regional Office for Asia and the Pacific.

Mithiya, D., Mandal, K., & Datta, L. (2018). Trend, pattern and determinants of crop diversification of small holders in West Bengal: A district-wise panel data analysis. *Journal of Development and Agricultural Economics, 10*(4), 110–119.

Mohammad, N. (1981). *Agricultural diversification and food security in the mountain ecosystem* (First ed., Vol. 3). Concept Publishing Company Pvt. Ltd. New Delhi.

Newby, H. (1983). The sociology of agriculture: Toward a new rural sociology. *Annual Review of Sociology, 9*(1), 67–81.

Petit, M., Barghouti, S., & Gnaegy, S. (1993). New challenges in agricultural technology transfer. *Agricultural Technologies for Market-led Development Opportunities in the 1990s, 23*, 5.

Pingali, P. L., & Rosegrant, M. W. (1995). Agricultural commercialization and diversification: Processes and policies. *Food Policy, 20*(3), 171–185.

Pitipunya, R. (1995). Determinants of crop diversification on paddy field: A case study of diversification to vegetables. *Kasetsart Journal (Social Science), 16*, 201–208.

Reinwald, L. T. (1949). *Some Aspects of Statistically Interpreting the Manufactural Functions of United States Cities* (Doctoral dissertation, Clark University)

Siddiqui, S. H., & Rahaman, H. (2016). Crop diversification in relation to time and space: a study from Malda district. *International Journal of Informative & Futuristic Research, 4*(2), 5133–5142

Shafi, S. P. (2008). *Diversified cropping pattern and agricultural development-a case study of Dadri block, Gautam Budh Nagar, UP* (Doctoral dissertation), Aligarh Muslim University.

Sharma, K. C. (2001). Crop diversification in Nepal. In *Crop diversification in the Asia-Pacific Region* (p. 81).

Shucksmith, D. M., & Smith, R. (1991). Farm household strategies and pluriactivity in upland Scotland. *Journal of Agricultural Economics, 42*(3), 340–353.

Simpson, E. H. (1949). Measurement of diversity. *Nature, 163*(4148), 688–688.

Singh, J. (1976). An agricultural geography of Haryana. In *An agricultural geography of Haryana* (pp. 187–190).

Tress, R. C. (1938). Unemployment and the diversification of industry 1. *The Manchester School, 9*(2), 140–152.

Vaishampayan, M. R., Nile, U. V., & Patil, D. Y. (2019). *Agriculture geography a study of cropping pattern in western Satpuda region* (pp. 140). Lulu publication, USA.

Van Luat, N. (2001). Crop diversification in Vietnam. In *Crop diversification in the Asia-Pacific region* (p. 147).

Vyas, V. S. (1996). Diversifcation in agriculture: Concept, rationale and approaches. *Indian Journal of Agricultural Economics, 51*(4), 636.

Yahya, T. M. B. T. (2001). Crop diversification in Malaysia. In *Crop diversification in the Asia-pacific region* (p. 64).

Chapter 5
Levels and Efficiency of Agricultural Development

The last two chapters on agricultural development and crop diversification are mainly based on secondary sources of data. In both chapters, performance of marginal land size is better than other land size groups which invalidate the notion of bigger land size results in better development. To check the development-diversification relation further, micro-level analysis has been done based on a field survey. In the present chapter, agricultural development along with crop diversification has been evaluated through households' surveys. The analyses include description of physical, socioeconomic, and demographic characteristics of the villages. This follows the soil nutritional status of sampled villages. Next to this, the relation of crop diversification with agricultural development is analyzed through index value. Subsequent to this, an attempt has been made to find out the factors of development and diversification through principal component analysis. In the last section, relative efficiency of agriculture with respect to input-output has been examined.

5.1 Importance of Field Work in Geography

One of the most important aspects of geography is fieldwork. The position of fieldwork and its clear role as a basic principle of the geographical educational experience is discussed by Claire Herrick in her article "Lost in the field: ensuring student learning in the 'threatened' geography field trip" (Area 2010). Herrick suggests that fieldwork can inspire a deep approach to learning and provide formative experiences for research. In research perspective, the real-time experience from the field ignites, correlates, and deduces important basis for what a researcher wondered (Hall and Healey 2005). Not only this, the field survey research and innovation have direct forward linkages with all policy intervention issues. In an academic point of view, household survey for investigation of agricultural situation in general and

© The Editor(s) (if applicable) and The Author(s), under exclusive license to   139
Springer Nature Switzerland AG 2021
H. Rahaman, *Diversified Cropping Pattern and Agricultural Development*,
https://doi.org/10.1007/978-3-030-55728-7_5

developmental situation in particular assists to realize the actual causes of why they are being in as such consequence. With this objective in mind, agriculture situation of selected villages has been assessed through household survey with the help of scheduled and questionnaires, interactive session, focus group discussion, and personal observation.

5.2 Brief Description of Sampled Villages

The household survey from 15 villages is conducted in the month of January and February, 2018. From each block, one village is selected for the study. The village having the maximum number of cultivators according to Census of India 2011 is selected for survey. The main context of study is agriculture; therefore, only cultivators get due advantage as a base of selection criterion. The purposive sampling technique has been used. A total of 30 samples have been selected from each village. From all villages, a total of 450 samples have been studied. The samples have collected in keeping view of fivefold land size classes. Apart from the socioeconomic information of respondents, other essential information related to agriculture and its development have been just at the end of the present section (Fig. 5.1 and Table 5.1).

Khutadaha
The village is situated ($25°10'28''$ N latitude and $88°26'08''$ E longitude) around 18 km east from community development block of Bamangola. The Bangladesh's border is just 3 km east from the village. According to 2011 Census, the said village is spread over an area of 694.4 ha and inhabited by 3979 people in 788 households. From the age group of 15–59, the total worker of the village is 1383, out of which 860 are cultivators. Around 52.23% soil of the village is moderately acidic, while electrical conductivity is 100%, i.e., normal. The organic carbon of the soil is low, that is, in 74.49% soil, phosphorus is very high (57.09%), and potassium is low (66.67%). In terms of micronutrients of soil, copper (77.73%) and iron (100%) levels are sufficient, but manganese (53.44%) and zinc (99.60%) are deficient.

Hatinda
The village is located ($25°24'54''$ N latitude and $87°59'23''$ E longitude) in Chanchal-I block which is 6 km west from block headquarter. It has 1454 households shared by 9248 populations and spread over an area of 433.9 ha. In total workers of 2600, 655 were cultivators as per 2011 Census. Soil pH is moderately alkaline (92.06%), electrical conductivity is normal (100%), organic carbon is low to very low (79.46%), the phosphorus content is very high (57.09%), and potassium of soil ranges from low (50%) to high (50%). Among micro soil nutrients, copper (84.1%), iron (96.83%), and manganese (76.19%) are in sufficient category, but soil suffers from zinc (90.48%) deficiency.

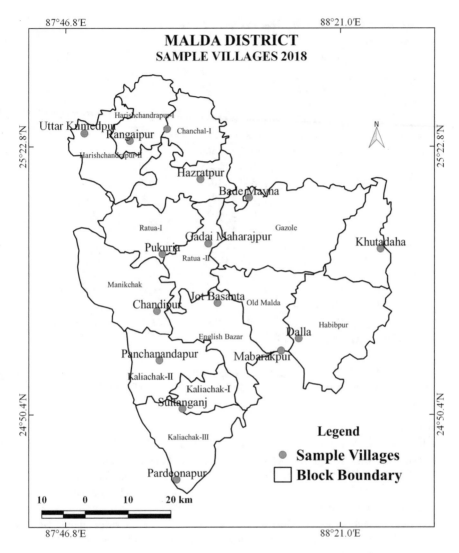

Fig. 5.1 Sample villages in Malda District 2018

Hazaratpur

Around 10 km south from block headquarter, Hazaratpur (25°18′50″ N latitude and 88°03′40″ E longitude) village is the home of 9248 populations under 1737 households which extends over 7458 ha of land. According to 2011 Census, the size of cultivators in the village is 1220 in 3244 total workers (aged 15–59 years). Soil quality in terms of pH index is slightly acidic (44.78%) to moderately alkaline (35.82%). Though electrical conductivity (100%) is normal, organic carbon content is very low (31.34%) to low (61.19%). The share of phosphorus (85.07%) and

Table 5.1 General information of selected sample villages 2011

(*Area in hectares*)

Block	Sampled villages	Total area	Total households	Total population	Total worker	Total cultivators
Bamangola	Khutadaha	694.4	788	3979	1383	860
Chanchal-I	Hatinda	433.9	1454	6177	2600	655
Chanchal-II	Hazaratpur	745.8	1737	9248	3244	1220
English Bazar	Jot Basanta	690.9	2388	11,438	3531	692
Gazole	Bade Mayna	1024.9	1368	6452	3135	1840
Habibpur	Dalla	483.4	1473	7177	3006	1152
Harishchandrapur-I	Rangaipur	340	1038	6002	1529	669
Harishchandrapur-II	Uttar Kumedpur	336.2	1041	6006	1649	843
Kaliachak-I	Sultanganj	642.2	4339	21,149	8696	910
Kaliachak-II	Panchanandapur	2301.9	5789	26,358	11,167	1302
Kaliachak-III	Par Deonapur	1012.1	2769	16,856	6531	1402
Manikchak	Chandipur	125.8	4067	16,017	6053	815
Old Malda	Mabarakpur	662.8	2268	9842	3456	817
Ratua-I	Gadai Maharajpur	1737.7	3144	15,023	5529	1479
Ratua-II	Pukhuria	846.3	4893	22,550	6769	808

Source: Village Directory, Census of India, 2011

potassium (54.55%) is very high and high, respectively. Copper (86.57%), iron (100%), and manganese (77.61%) are sufficiently good, but 92.54% soil suffer from zinc deficiency.

Jot Basanta
Jot Basanta village is located at 25°04′18″ N latitude and 88°05′25″ E longitude and inhabited by 11,438 populations in 2388 households and spread over an area of 690.9 ha. Out of 3531 total workforces, 692 are cultivators from main and marginal categories. Soil pH is moderately alkaline (75.00%), electrical conductivity is normal (100%), organic carbon is low (53.13%) to medium (31.25%), and phosphorus (59.38%) and potassium (75.0%) vary from very high to high, respectively. Except zinc (87.50%), other micronutrients such as copper (96.00%), iron (100%), and manganese are sufficient in soil (Fig. 5.2).

Bade Mayna
This village (25°17′23″ N latitude and 88°09′38″ E longitude) is located about 17 km northwest from Gazole blocks' headquarter. According to 2011 Census, Bade Mayna village spread over an area of 1024.9 ha comprises 1368 households and sustenance 6452 people. Out of 3135 workers, 1840 are cultivators. The soil of the village is moderately acidic (80.80%) and having normal (100%) conductivity. Organic carbon reports low (48.0%) to medium (27.20%). Phosphorus ranges from low (32.00%) to medium (36.80%). Potassium contains very low (9.52%) to low (90.48%). Copper (81.60%), iron (100.0%), and manganese (72.2%) nutrients are sufficient, while zinc deficiency is 99.20% in soil.

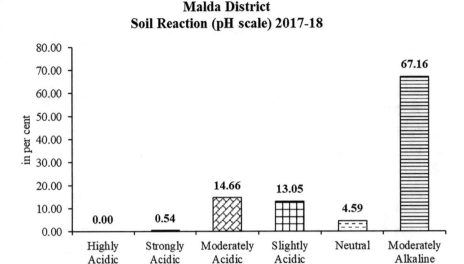

Fig. 5.2 Soil reaction (pH scale) in Malda District 2017–2018. (Source: based on soil health card report 2017–2018)

Dalla

Dalla is one of the villages in Habibpur block which is located (25°04′15″ N latitude and 88°25′06″ E longitude) around 17 km east from blocks' headquarter. The total population of the village is 7177 accommodated in 1473 households. The village extends over an area of 483.4 ha. Out of 3006 total workers, 1152 are cultivators. The soil of the village is moderately alkaline (66.19%), and conductivity is normal (100%). The organic carbon of soil ranges from low (71.94%) to medium (20.14%). Phosphorus (66.91%) and potassium (68.97%) report high in soil. Copper (97.12%), iron (100%), and manganese (69.06%) are sufficient, while zinc is confirmed as 100% deficient in soil.

Rangaipur

This village is located (25°23′31″ N latitude and 87°54′48″ E longitude) in Harishchandrapur-I block which is 5 km east from block headquarter and 100 m away from the main road. Rangaipur is populated by 6002 people in 1038 houses and extends over 340 hectares' area. The total cultivators in this village are 669, and total workers are 1529. Moderately alkaline soil is 66.67% and 20.51% soil reported as slightly acidic. The electrical conductivity of soil is normal (100%). About 66.67% soil holds low organic carbon, and 23.08% remain in medium range. Phosphorus and potassium report from low to high. Zinc (100%) and manganese (51.28%) are deficient in soil. The iron and copper are sufficient in soil up to 79.49% and 100%, respectively (Fig. 5.3).

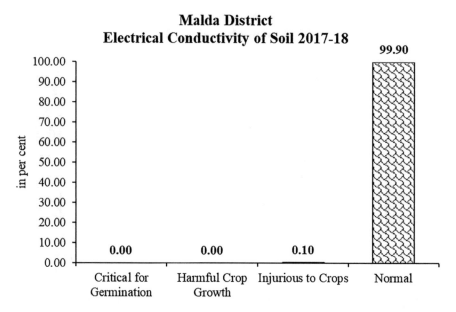

Fig. 5.3 Electrical conductivity of soil in Malda District 2017–2018. (Source: based on soil health card report 2017–2018)

Uttar Kumedpur

The village is spread over an area of 336.2 ha. This village is located between 25°24′24″ N latitude and 87°49′05″ E longitude and inhabited by 6006 peoples in 1041 households. Out of 1649 total workers, 843 are cultivators. Moderately alkaline soil locates 70.91%, while soil conductivity is normal (100%). About 61.82% soil is deficient in organic carbon. Soil holds very high (58.18%) to high (66.67%) phosphorus and potassium, respectively. Cooper (80.0%), iron (100%), and manganese (61.82%) have been found sufficient, while zinc notes deficient (70.91%) in soil.

Sultanganj

This village is situated 1 km away from the main road. It is located at 24°52′05″ N latitude and 88°01′52″ E longitude in Kaliachak-I block. The village is the home of 910 cultivators and 8696 workers. The village has 21,149 populations under 4339 households and spread over 642.2 ha of land. About 95.71% soil of the village is moderately alkaline, and electrical conductivity is also normal up to 100%. Organic carbon of the soil varies from low (40.71%) to medium (48. 57%). Phosphorus quantity ranges from medium (30.36%), high (25.71%), and very high (42.14%). Potassium (63.21%) reports medium in range. Copper (80.51%), iron (90.71%), and manganese (65.34%) are sufficient, and 88.93% soil report zinc deficiency (Fig. 5.4).

Panchanandapur

It is situated (24°56′50″ N and 87°58′34″ E) in Kaliachak-II which is 12 km west from blocks' headquarters. The village is the home of 26,358 populations and lives in 5789 houses. The village covers an area of 2301.9 ha. In total of 11,167 workers,

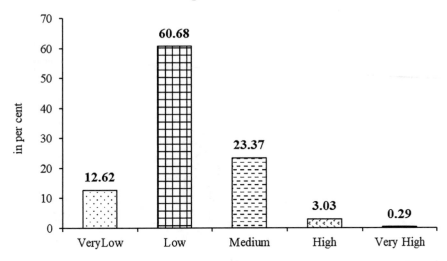

Fig. 5.4 Organic carbon in soil in Malda District 2017–2018. (Source: based on soil health card report 2017–2018)

1302 are cultivators. Soil pH of said village is moderately alkaline (up to 88.87%) and 100% soil reports as normal. Very low and low organic carbon in soil comprises around 81.55%. Phosphorus contains varies from medium (55.13%) to high 22.24 (22.24%). Soil potassium ranges from 62.28% to 30.75% in the medium and high category. Except zinc (91.30%), copper, iron, and manganese are sufficient up to 88.97%, 99.81%, and 69.25%, respectively.

Par Deonapur
The village Par Deonapur is located in Kaliachak-III block. It has a population of 16,856 housed in a 2769 family. It spread over an area of 1012.1 ha. It lies on 24°41′39″ north latitude and 87°59′50″ east longitudes. The electrical conductivity of soil shows normal (100%) condition. The 100% soil is tested as moderately alkaline, and 80.00% soil is deficient from low organic carbon. The soil of the village contains very high phosphate (51.2%) and moderately potash (69.79%). The micronutrients such as copper (100%), iron (100%), and manganese (56.67%) are in sufficient category, but zinc remains deficient (100%) in soil (Fig. 5.5).

Chandipur
It is situated (25°02′56″ N latitude and 87°58′08″ E longitude) in Manikchak block. The village expands over an area of 125.8 ha and supports 4067 households. It has 16,017 populations in which 6053 are workers and 815 are cultivators. Soil pH of the village is moderately alkaline (34%), slightly acidic (32%), and moderately acidic (26%). The soil has 22% organic carbon in a very low category, and 16% found in medium class. Phosphorus and potassium of soil depict 92% (very high

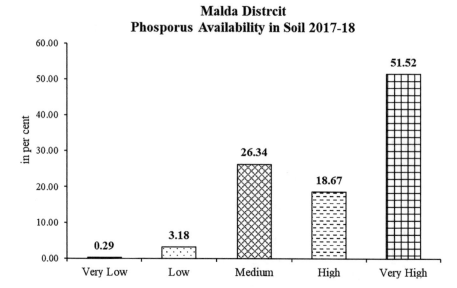

Fig. 5.5 Phosphorus availability in soil in Malda District 2017–2018. (Source: based on soil health card report 2017–2018)

category) and 46.67% (high), respectively. Copper (90%), iron (100%), and manganese (56.67%) nutrients are sufficient, and zinc remain deficient in 100% soil.

Mabarakpur
It is situated about 11 km away from the block headquarter of Old Malda block. Mabarakpur village (24°58′06″ N latitude and 88°13′39″ E longitude) is a home of 9842 people in which 817 are cultivators in a total of 3456 workers. It has 2268 households which are extended over 662.8 ha' area. Soil pH of the village is mostly moderately alkaline (91.30%), and electrical conductivity is normal (98.55%). Organic carbon ranges from low (19.57%) to very low (65.22%). Phosphorus and potassium of soil report as 70.29% and 62.67% in very high and low categories, respectively. The micronutrients such as copper (96.38%), iron (100%), and manganese (72.42%) are sufficient, but 97.10% soil is zinc deficient.

Gadai Maharajpur
It is situated (25°11′00″ N latitude and 88°04′31″ E longitude) in Ratua-I block which is 2 km far from block headquarter. The village is the home of 15,023 inhabitants. It has 1479 cultivators and 5529 workers. The village occupies an area of 1737.7 ha and houses 3144 families. The soil of the village is moderately alkaline (87.06%). Soil of the village is normal (100%) in terms of electrical conductivity. Organic carbon of soil varies from low (51.76%) and medium (32.94%). The share of phosphates (70.59%) and potassium (76.19%) has been categorized under very high and high, respectively. Among micronutrients, copper (88.24%), iron (100%), and manganese (76.47%) are sufficient in soil, but zinc (75.29%) sits deficient (Fig. 5.6).

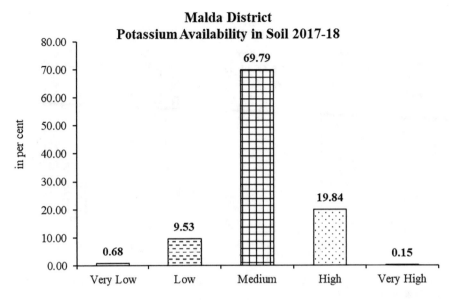

Fig. 5.6 Potassium availability in soil in Malda District 2017–2018. (Source: based on soil health card report 2017–2018)

Pukhuria

This village is located (25°07′54″ N latitude and 88°01′54″ E longitude) in Ratua-II block. It is spread over an area of 846.7 ha and housed by 4893 families (Village Directory, Census of India, 2011). It has 22,550 populations. Soil pH of the village is moderately alkaline (48.04%), slightly acidic (25.14%), and moderately acidic (17.32%). The electrical conductivity of soil is normal (100%). The share of organic carbon ranges from very low (13.48%), low (51.69), and medium (30.34%). Phosphorus in soil is 96.06%, while potassium is 88.27% under the medium category. Data on copper and manganese are not available. Around 50.57% soil is iron scarce, and 93.85% is zinc deficient (Fig. 5.7).

5.3   Soil Nutrition Index in Sampled Villages

After realizing the importance of soil nutrients for growth and production of plants, an attempt has been made to calculate their composite index under micro, macro, and overall soil nutrient index at the village level in the district. Macro soil nutrients comprise nitrogen (N), phosphate (P), potassium (K), and organic carbon (OC). Figure 5.8 shows that in terms of macronutrients, Pardeonapur village found better index value followed by Dalla in the district. The least index value observed for Mabarakpur village hence the quality of soil in said village is low as compared to the rest of the villages. The other villages represent index values below 0.500 but above 0.350 in the district .

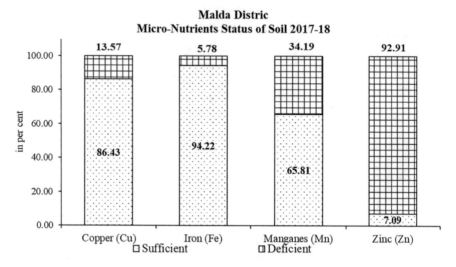

Fig. 5.7 Micro-nutrients status of soil in Malda District 2017–2018. (Source: based on soil health card report 2017–2018)

The spatial pattern of macro soil nutrient is presented in Fig. 5.8. Figure 5.8 shows that index value below the first quartile has been found in two villages, namely, Bade Mayna and Chandipur, while index value of the above third quartile is observed in Dalla and Par Deonapur villages. The remaining villages lie between 0. 409 and 0.492 index value which signifies moderate nutrient index.

Apart from macro soil nutrients, plants also need small quantities of iron, manganese, zinc, copper, boron, and molybdenum, known as trace elements or micronutrients. The role these nutrients play in plant growth is complex and essential. Among the villages, Pukhuria represents the healthiest in terms of micro soil nutrients. Pardeonapur, Khutadaha, Rangaipur, and Chandipur are also found with more than 0.500 index value. Panchanandapur and Hatinda have scored equally lowest among the villages in the district. The spatial pattern of micro soil nutrients (Fig. 5.9) reveals that five villages lie above the upper quartile and five villages come under below the lower quartile. The remaining village is represented in medium quartile index. Figure 5.9 also suggests that the micro soil nutrients in the district lacks in most of the villages.

The overall soil nutrients index reports that Par Deonapur (0.600) village is the healthiest followed by Dalla (0500) and Pukhuria (0.490) village in the district. Mabarakpur with index value 0.350 is the most nutrients deficient village. Bade Mayna (0.397), Sultanganj (0.413), and Panchanandapur (0.417) sit successively lower index value. Except in three villages, namely, Pardeonapur, Mabarakpur, and Bade Mayna, other villages of the district do not show significant difference in overall nutrients index. It is noted that micronutrients in soil are significantly low in five villages, but the overall index does not seem low due to better macronutrients in soil.

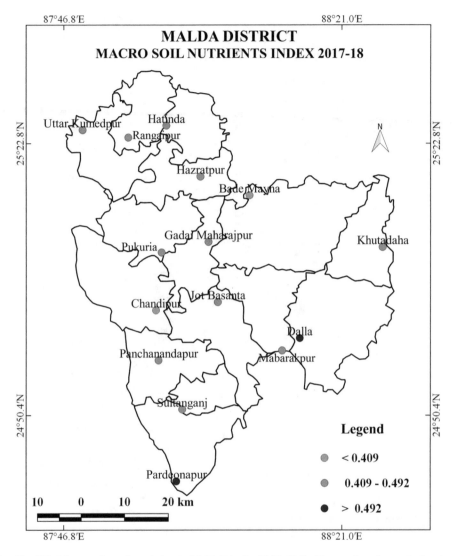

Fig. 5.8 Macro soil nutrients index in Malda District 2017–2018. (Source: based on calculated value from Annexure 5a, b)

The spatial pattern of overall soil nutrients has been presented in Fig. 5.10. It depicts that four villages scored more than upper quartile, seven are in moderate range, and four villages come under the lowest quartile range. Overall soil nutrients for Sultanganj, Panchanandapur, Chandipur, and Hatinda are below the lower quartile signifying low soil quality with respect to other villages in the district.

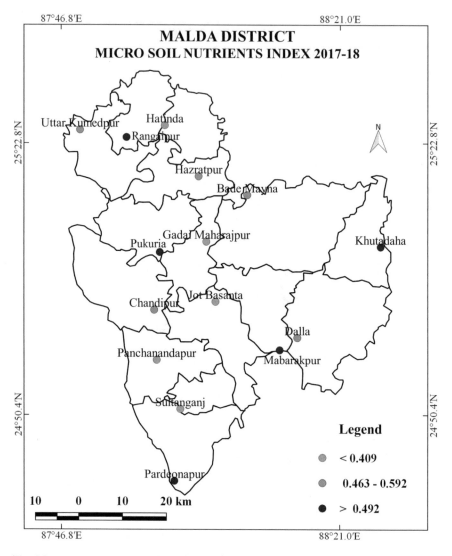

Fig. 5.9 Micro soil nutrients index in Malda District 2017–2018. (Source: based on calculated value from Annexure 5a, b)

5.4 Agricultural Circumstances in Sampled Villages

In present analysis on agriculture development, 32 variables have been selected for the study which is further classified into 8 indicators based upon the literature survey. The spatial information on each aspect of agriculture development has presented land size class category wise on the total sample of 450, and their spatial analysis has been explained based upon field survey data (attached in Annexures 3c–3g) (Table 5.2).

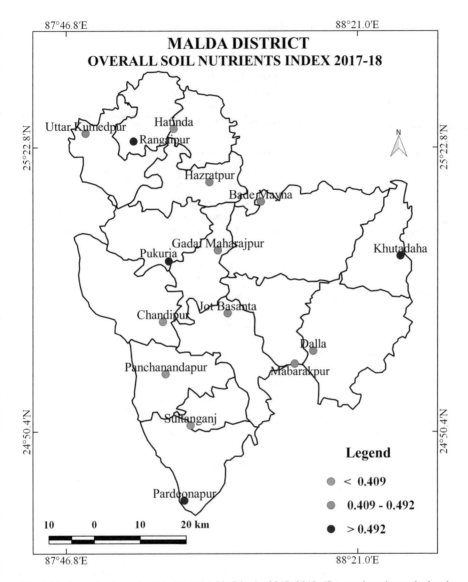

Fig. 5.10 Overall soil nutrients index in Malda District 2017–2018. (Source: based on calculated value from Annexure 5a, b)

Land Size Class Wise Respondents

Out of 450 total respondents, 360 are from marginal land category (< 1.0 ha), 45 from small land size (1–1.99 ha), 30 from semi-medium (2–3.99 ha), and 15 from medium land size class (4–9.99 ha). There is no large farmer found in selected villages; therefore, the large farmer category is omitted from analysis.

Table 5.2 List of indicators/variables of agriculture development and crop diversification

Indicators	Variables	Description of variables
Land	**X1**	Number of households by land size classes
	X2	Average landholding/land size
	X3	Percentage of wholly own and self-operated area
	X4	Percentage of wholly leased-in area
	X5	Percentage in the total area located entirely within the village of residence
Irrigation	**X6**	Percentage of net irrigated area in total GCA
	X7	Number of electric well and tube well per 5 ha of GCA
	X8	Number of diesel well and tube well in 10 ha of GCA
Seeds	**X9**	Percentage of households used certified seeds
	X10	Percentage of households used HYV seeds
Income and credits	**X11**	Number of HH earn 3000 per month from agriculture
	X12	Percentage of HH took institutional credit
	X13	Percentage of HH took credit from money lender
	X14	Percentage of HH have KCC
Machinery	**X15**	Number of hand-operated machine used in per hectare of GCA
	X16	Percentage of HH use power-operated machines
	X17	Percentage of HH used tractor and harvester
soil and fertilizer	**X18**	Percentage of HH have soil health card
	X19	Percentage of the area treated with insecticide and pesticides in GCA
	X20	Percentage of the area treated with chemical fertilizers
	X21	Amount of NPK complex used in 1 ha GCA (kg)
Market and cold storage	**X22**	Percentage of products sold directly in the market
	X23	Percentage of product sold directly from home and farm
	X24	Percentage of HH benefitted from cold storage facility
Socioeconomic	**X25**	Farming experience (in number of years)
	X26	Percentage of main worker in total worker
	X27	Percentage of marginal worker in total worker
	X28	Literacy
	X29	Number of farmers ever been participated in capacity building program
	X30	Number of farmers ever been participated in farmer agitation
	X31	Average size of HH
	X32	The average age of household 15–60 years age group

According to 2015–2016 Agriculture Census, 83.40% farmer belongs to marginal and followed by 12.65 in small, 3.66% from semi-medium, and 0.29% from medium land size category. Here, the proportion of respondents is selected according to the Agriculture Census finding the share of farmers. It is clear that the district is going through marginalization of landholding due to increasing population pressure and nucleation in family size (Fig. 5.11).

Malda District
Number of Samples by Land Size Classes 2018

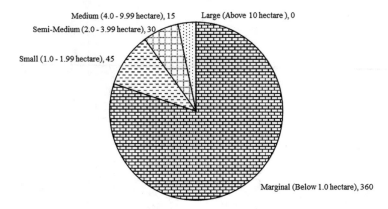

Fig. 5.11 Number of samples by land size classes from Malda District 2018. (Source: based on field survey 2018 (see Annexures 5c–5g))

Average Size of Land Holding
From all sampled villages, the average land size in the district is 0.43 ha per household. In marginal land size category, the average land size is 0.35 ha, 1.29 ha in small, 2.21 ha in semi-medium, and 4.66 ha in the medium category (Fig. 5.12). The size of land holdings in marginal category is low due to the nuclear family and population explosion. In the recent past, migration has been an important cause in nucleated family formation from the joint family system.

Wholly Owned and Self-Operated Area
The wholly owned and self-operated cultivable land in marginal land size class is high due to small parcels of land which is the only asset of their livelihood. Due to limited parcels of land in marginal and small land size farmers, they do not lease out land rather they leased in from semi-medium and medium land size holders because the intensive subsistence cum low commercial agriculture is only a boost of their income. The semi-medium and medium land size farmers generally leased out their land due to decreasing profit from agriculture (Fig. 5.13).

Wholly Leased-in Area
Though agriculture is not profitable to marginal and small land size farmers, they take leased-in land because of cereals for sake of family consumption. Under the marginal land size category, around 4.31 land is reported as leased-in area. The small land size farmer takes around 2.27% land on leased (Fig. 5.14). In total cultivable land, the all land classes group stand third with 3.23% leased-in area. Aged and poor farmers prefer to take land on lease where fixed produce and fixed money are generally offered as terms of leasing.

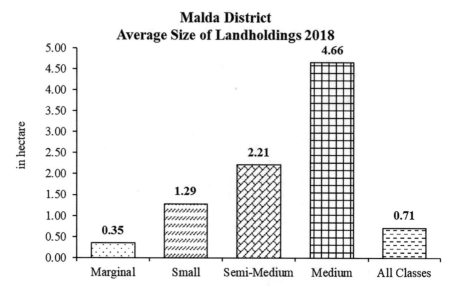

Fig. 5.12 Average size of landholdings in Malda District 2018. (Source: based on field survey 2018 (see Annexures 5c–5g))

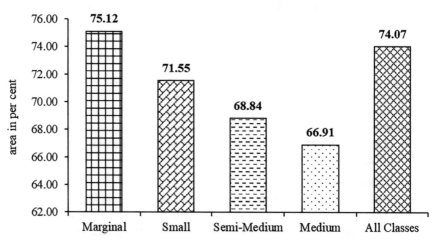

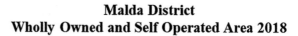

Fig. 5.13 Wholly owned and self-operated area in Malda District 2018. (Source: based on field survey 2018 (see Annexures 5c–5g))

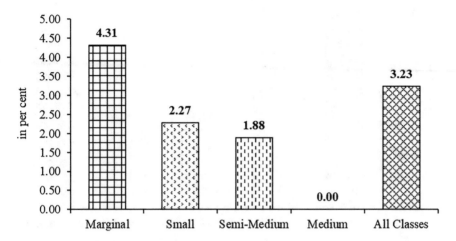

Fig. 5.14 Leased-in area in Malda District 2018. (Source: based on field survey 2018 (see Annexures 5c–5g))

Location of Net Sown Area

Location of the net sown area is an important aspect in the determination of crops to be grown. If the land area is away from the village (distant), farmers generally prefer plantation crops, fruits, and cash crops. On the other hand, cereals, vegetables, oilseeds, and floriculture required regular look after, and therefore, these crops are generally produced near to the house (Fig. 5.15).

A total of 89.05% land is located within villages which belong to marginal farmers. The small (84.72%), semi-medium (81.68%), and medium (81.48%) are having widespread location due to bigger land size.

Net Irrigated Area

The marginal land size farmers are having low net irrigated area due to limited sources of irrigation such as pump set and tube well. More than 60% cultivated area sources irrigation from monsoon rainfall mainly between May and July. Several wetlands, ponds, beels, and groundwater are the major sources of irrigation in the remaining months of the year (Fig. 5.16).

Electric Wells and Tube Wells

Intensive subsistence cum commercial agriculture in small, semi-medium, and medium land size categories requires ensured and regular irrigation sources; therefore, the number of electric-generated wells and tube wells which are cost and time efficient found higher in said land size categories. In sampled villages, though electric wells and tube well are found in abundances, lack of regular and timely electricity caused hindrance in use of such irrigation measures. In per 5 ha GCA of land,

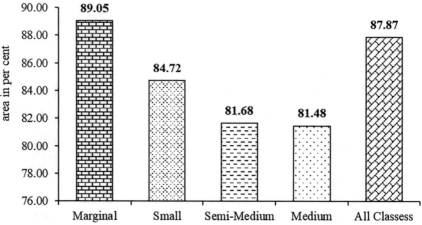

Fig. 5.15 Net sown area located within village in Malda District 2018. (Source: based on field survey 2018 (see Annexures 5c–5g))

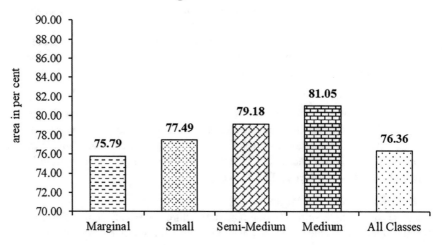

Fig. 5.16 Net irrigated area in Malda District 2018. (Source: based on field survey 2018 (see Annexures 5c–5g))

Malda District
Number of Electric Well and Tubewell in per Five Hectare of GCA 2018

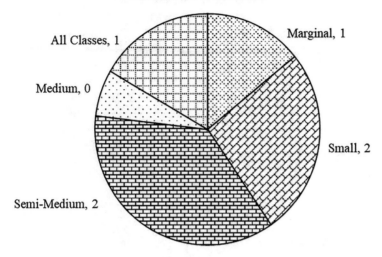

Fig. 5.17 Number of electric well and tube well in per 5 ha of GCA in Malda District 2018. (Source: based on field survey 2018 (see Annexures 5c–5g))

marginal farmer (01) secures less wells and tube wells as compared to small (02), semi-medium (02), medium (02), and all land class (02) in the district (Fig. 5.17).

Diesel-Generated Wells and Tube wells

Although diesel-operated machines are costlier than electric-generated well and tube wells, diesel-operated machines are more in number than the electric because of lack in ensured and timely electricity. One-time factor cost of electric machines is much higher than diesel-operated machines, and that is why farmers preferred diesel-generated machines for timely irrigation. Figure 5.18 shows that except medium land size group, other land size classes are having three wells and tube well in per 5 ha of GCA in district.

Certified Seeds

Seed certification is a legally sanctioned system for quality control of seed multiplication and production. The purpose of seed certification is to assure quality of seeds and genetic purity. Certified seed is the starting point of a successful crop as well as an important risk management tool. Figure 5.19 reports that the small land size farmers at the higher side in use of certified seeds because of intensive cultivation.

HYV Seeds

The HYV seed promises better productivity in the presence of assured irrigation infrastructure, insecticides, pesticides, and fertilizers. More than 77% seeds belong to the HYV category is mainly in cereals, vegetables, and oilseed category. The HYV seeds in fruits, fiber, and pulses are around 50% reported by villagers. Marginal and small farmers dominantly cultivate cereals which are intensive in

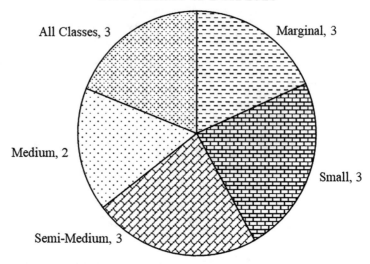

Fig. 5.18 Number of diesel operated well and tube well in per 5 ha of GCA in Malda District 2018. (Source: based on field survey 2018 (see Annexures 5c–5g))

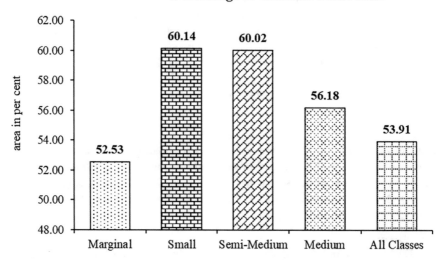

Fig. 5.19 Percentage of certified seeds used in Malda District 2018. (Source: based on field survey 2018 (see Annexures 5c–5g))

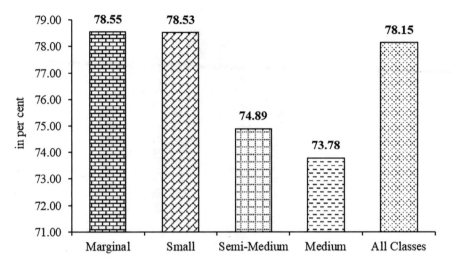

Fig. 5.20 SPercentage of HYV seeds used in Malda District 2018. (Source: based on field survey 2018 (see Annexures 5c–5g))

nature, and therefore, overall use of hybrid seeds is more than 78% in these land size categories. Due to limited and small parcels of land, poor farmers use HYV seeds for maximum output.

Monthly Income from Agriculture
Out of 360 marginal farmers, around 20% reports that they earn more than 3000 rupees per month exclusively from agriculture. This income is mainly from vegetable and spice crop cultivation. The small, semi-medium, and medium land size classes are having big land size and do practice intensive subsistence cum commercial agriculture; therefore, their monthly returns from agriculture are higher than marginal group. The decreasing agricultural prices, volatile market, and natural calamities have been reducing monthly earnings; therefore a very small number of farmers replied that they have low monthly return from agriculture.

Credit Facility
Less than 15% of total households (450) avails institutional credit from different banks and government institutions. The nature of credit is small, i.e., for 1 year in 90% cases. As a mortgage, the paper of owned land is considered by banks; therefore, medium and semi-medium land size holders get prior preference over marginal and small land size holders (Fig. 5.22).

Malda District
Number Households Gain Monthly Income from
Agriculture > 3000 Rupees 2018

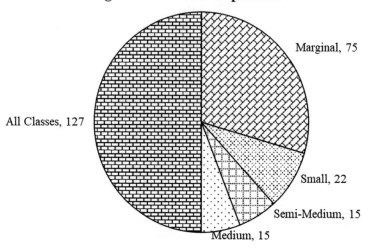

Fig. 5.21 Number of households gain monthly income from agriculture > 3000 rupees in Malda District 2018. (Source: based on field survey 2018 (see Annexures 5c–5g))

Malda District
Number of HH Avail Institutional Credit Facility 2018

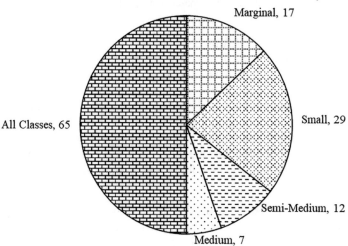

Fig. 5.22 Number of HH that availed institutional credit facility in Malda District 2018. (Source: based on field survey 2018 (see Annexures 5c–5g))

Malda District
Number of HH Avail Credit from Money Lender 2018

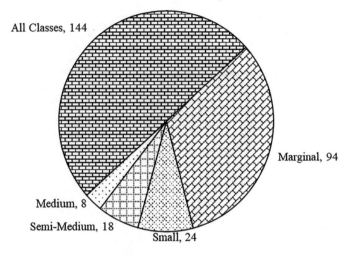

Fig. 5.23 Number of HH that availed credit from money lender in Malda District 2018. (Source: based on field survey 2018 (see Annexures 5c–5g))

Credit from Money Lender

The credit in the form of dues for seeds, fertilizers, other agriculture implements, and even cash from shopkeepers and money lenders often charge high interest. Because of limited institutional credit facilities, they are submissive to take loans and credits from shopkeepers and village money lenders even at a higher interest rate. Figure 5.23 depicts that 32% in total respondents are credited from money lenders in which shares of marginal (36.11%), small (53.33%), semi-medium (60%), and medium (53.33%) are noted above 35%.

Kisan Credit Card (KCC)

With the aim of quick, timely, and affordable credits for the farmer, Reserve Bank of India and NABARD jointly lunched this scheme in 1998. The KCC is offered by the regional rural bank, cooperative banks, and public sector banks. Though around 20% of the farmer has received KCC well before, they have not availed credit facilities, yet, it means having KCC not guaranteed accessibility of agricultural credit (Fig. 5.24).

Hand- and Animal-Operated Machineries

In intensive subsistence farming, the extensive use of agricultural implements is difficult and expensive. And therefore, traditional methods of cultivation persist from ages long. Figure 5.25 shows that the number of hand- and animal-operated machinery is higher in marginal and small land size class farmers due to the small and fragmented parcel of land.

Malda District
Number of HH have Kisan Credit Card 2018

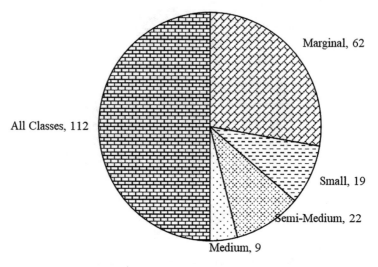

Fig. 5.24 Number of HH that have Kisan credit card in Malda District 2018. (Source: based on field survey 2018 (see Annexures 5c–5g))

Malda District
Number of Hand and Animal Operated Machinaries
Per Hectare GCA 2018

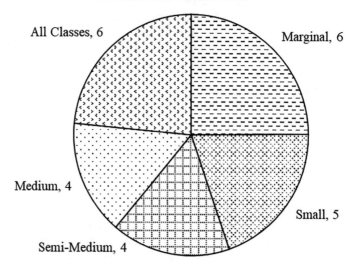

Fig. 5.25 Number of hand- and animal-operated machineries used in per hectare GCA in Malda District 2018. (Source: based on field survey 2018 (see Annexures 5c–5g))

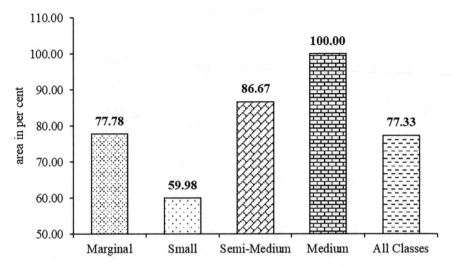

Malda District
Percentage of HH Use Power Opertaed Machines 2018

Fig. 5.26 Percentage of HH that use power operated machines in Malda District 2018. (Source: based on field survey 2018 (see Annexures 5c–5g))

Power-Operated Machines

The access of power-operated machines such as motor pump sets, frieze, motorbike, thresher, and others is 100% in medium and 86.67% in semi-medium land size categories. The use of power-operated machines under said categories is higher than rest groups due to bigger land size. The increasing electrification and farmer income contribute positively to the total increase of power-operated machines in the district. The ease of operating multipurpose machines is lucrative to farmers for a long time, and therefore, higher-income groups such as medium land size farmers and semi-medium income groups purchase and use these for agricultural purposes (Fig. 5.26).

Tractor, Harvester, and Sprayer

Because of small parcels of land in marginal and small land classes, they have often faced difficulties with mechanized farming. However, the application of modern machinery is time efficient but not the cost of efficiency. Therefore, marginal and small landholders represent low share in power-operated machineries (Fig. 5.27).

Soil Health Card

Soil health card scheme is launched by the Government of India which aims at crop-wise recommendations of nutrients and fertilizer that have been required to individual farms to improve productivity through judicious used of inputs. Against the soil testing, the authority issued a health card which recommended required nutrition ratio for soil health. The field survey shows that less than 10% farmers have soil health card in the district (Fig. 5.28).

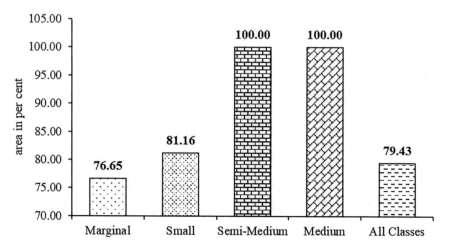

Malda District
Percentage of GCA used Tractor, Harvester and
Sprayer 2018

Fig. 5.27 Percentage of GCA that used tractor, harvester and sprayer in Malda District 2018. (Source: based on field survey 2018 (see Annexures 5c–5g))

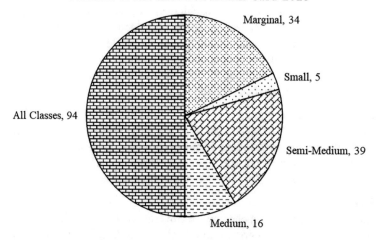

Malda District
Number of HH have Soil Health Card 2018

Fig. 5.28 Number of HH that have soil health card in Malda District 2018. (Source: based on field survey 2018 (see Annexures 5c–5g))

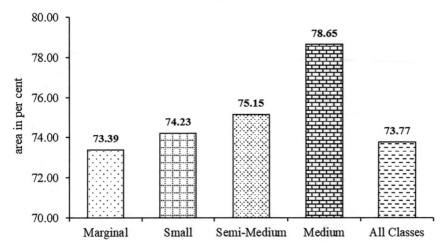

Fig. 5.29 Percentage of GCA treated with insecticide and pesticides in Malda District 2018. (Source: based on field survey 2018 (see Annexures 5c–5g))

Insecticides and Pesticides

The crop disease especially vegetables and fruits is increasing day by day. In such a case, the proper doses of insecticides and pesticides are viable option for better productivity. Though cereals and fiber are affected by insects and pests in more, it is lesser than pulses and oilseeds. Figure 5.29 shows that marginal (89.36%) and small (88.32%) land size farmers have used more insecticides and pesticides for good productivity than semi-medium and medium farmers in the district.

Chemical Fertilizers

The inorganic and synthetic fertilizers are often used to supply plant nutrition for better plants growth and productivity. Three types of chemical fertilizers such as nitrogenous, phosphoric, and potash are abundantly used. The percentage of GCA treated with chemical fertilizers shows that marginal (90.26%) and small (81.23%) land size farmers use heavy doses of fertilizers than other land size classes (Fig. 5.30).

Amount of NPK Mixture

The marginal farmer uses NPK mixture more than 500 kg/ha which is maximum among land size groups. The surveyed villages bestow with two to four crops on the same piece of land in a year dominated under cereals and vegetables. It is reported that the maximum fertilizers used all time for vegetable crops more than other crop categories.

Malda District
Percentage of GCA Treated with Chemical Fertilisers
2018

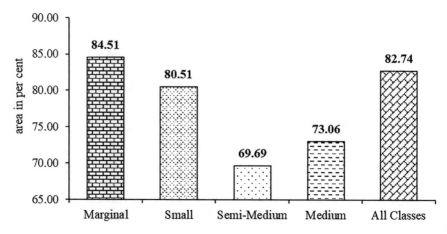

Fig. 5.30 Percentage of GCA treated with chemical fertilizers in Malda District 2018. (Source: based on field survey 2018 (see Annexures 5c–5g))

Malda District
Amount of NPK Mixture use Kg/Per Hectare 2018

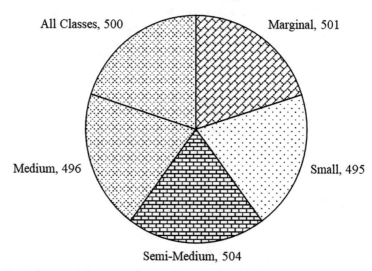

Fig. 5.31 Amount of NPK mixture used (kg/ha) in Malda District 2018. (Source: based on field survey 2018 (see Annexures 5c–5g))

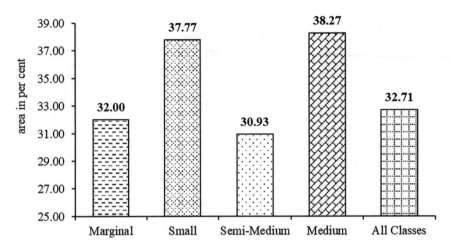

Fig. 5.32 Percentage of agro-products sold directly to markets in Malda District 2018. (Source: based on field survey 2018 (see Annexures 5c–5g))

Market Opportunities

In total sold, around 32% crop is directly sold in the market by marginal land farmers. In the marginal and small land size category, main crops sold in the market are vegetables followed by fruits and cereals. In total crops, around 37% produce is sold directly in market by small land size farmers, while 30% by semi-medium and 38% by medium land size farmers in district. The accessibility of markets to medium land class farmers is highest because the larger percentage of crops produces for sale (Figs. 5.31 and 5.32).

Crops Sold Directly from Home/Field

The marginal (68%) and semi-medium (69.07%) land size farmers sold crops directly from home or field are cereals, oilseed, fruits, and fiber due to less accessibility of market. Because of the limited number of fair price shops, regulated markets, and registered shops, the farmers are bound to sell crops from home or field. It results in low profit (Fig. 5.33).

Cold Storage

The district has only two cold storages for potatoes in Samsi (under Ratua-II block) and Adina (in Old Malda block). Out of 450 farmers, only 16 avail cold storage facilities. It is a long-time demand of farmers to increase the number of cold storages, and it's for other crops such as fruits and other vegetables. To avoid the risk and increase farm income, it is essential to increase the number cold storages for different land size farmers in the district.

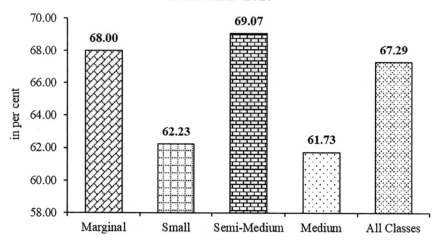

Fig. 5.33 Percentage of agro-products sold directly from home/field in Malda District 2018. (Source: based on field survey 2018 (see Annexures 5c–5g))

Farming Experience

The average year of farming experience in small (35 years) and marginal (33 years) land size farmers is more than 30 years. It indicates that the aged-old cultivation is being practiced in the district. Although the farming experience is also necessary for profitable agriculture, given infrastructure must be progressive technologically and institutionally (Saran 2003 cited by Siddiqui 2007).

Main Cultivator

A farmer is said to be the main cultivator only when one engages in agricultural activities for more than 6 months in a year (Census of India). Because of large land size, semi-medium (100%) and medium (100%) land size farmers engage throughout the years in agricultural practices, and therefore, the respondents in these categories have reported as main cultivators. The share of main cultivators in marginal land size category (44.2%) is low compared to small farmers mainly small land base (Figs. 5.34, 5.35, 5.36, 5.37, and 5.38).

Marginal Cultivator

According to Census of India, a farmer works less than 6 months in a year, and then one shall be categorized as a marginal cultivator. Due to lack of owned and self-operated land, farmers migrate toward cities. Therefore, the share of marginal cultivators in marginal land size category is the maximum (55.80%) compared to other land size classes. All land size class category shows that 39.56% cultivators belong to the marginal farming category in the district.

Malda District
Number of HH Avail Cold Storage Facility 2018

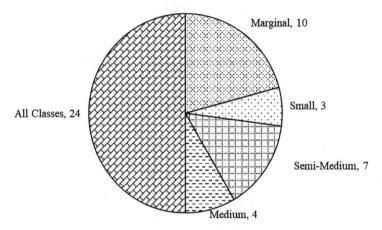

Fig. 5.34 Number of HH availed cold storage facility in Malda District 2018. (Source: based on field survey 2018 (see Annexures 5c–5g))

Malda District
Average Farming Experience (year) 2018

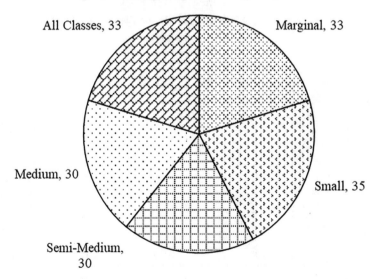

Fig. 5.35 Average years of farming experience in Malda District 2018. (Source: based on field survey 2018 (see Annexures 5c–5 g))

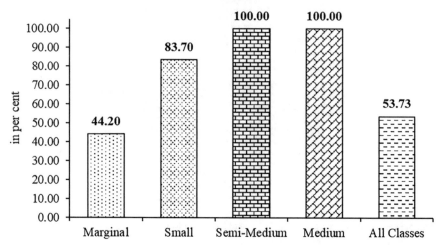

Fig. 5.36 Percentage of main cultivators in total cultivator in Malda District 2018. (Source: based on field survey 2018 (see Annexures 5c–5g))

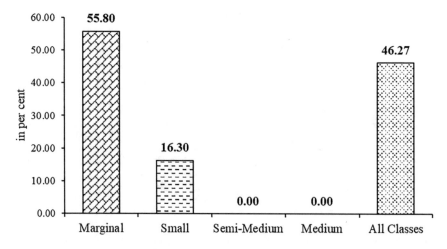

Fig. 5.37 Percentage of marginal cultivators in total cultivator in Malda District 2018. (Source: based on field survey 2018 (see Annexures 5c–5g))

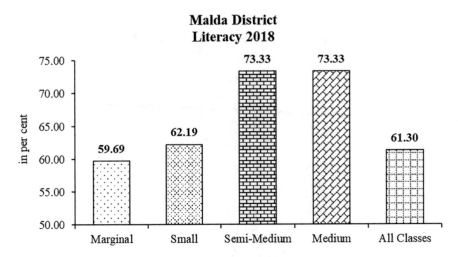

Fig. 5.38 Literacy in Malda District 2018. (Source: based on field survey 2018 (see Annexures 5c–5g))

Literacy

Marginal farmer remains in the lower tail of literacy due to incidence of poverty and child labor (<14-year age). The chain of child and adult migration toward cities for jobs has increased the dropout rate. The low income is the cause of out-migration; hence school-age children became child laborers as responded by famers. The semi-medium (73.33%) and medium (73.33%) land size farmers have been enjoying better literacy because of ensured income sources.

Capacity Building Program

It is important to update with technological know-how and make them aware to respond to the changing agricultural environment through extensions support, training on techniques, and sharing of best practices. For this, importance of capacity building program is highly felt. It helps farmers to adopt agricultural innovation in changing time. Figure 5.39 reveals that only 9.33% farmers of the district have attained any capacity building program in their lifetime.

Farmer Agitation

Frequent farmer agitation is an ill result caused from dissatisfaction of government policies and programs of agriculture. To be very specific, farmer agitation in the district is mainly because of reducing government incentive in agriculture, sudden price falling, reducing minimum support price (MSP), credit facilities, and non-fulfilment of announced promises which have been reported by farmers. The participation of marginal farmers in agitation was more in number due to reducing price and government incentive which caused huge losses of said category cultivator (Fig. 5.40). Out of 450 respondents, just 50 respondents reported as participants in agriculture related agitation ever in their lime time. Unpleasant impact of land disintegration, land tenancy policy of government, and MSP failing have been the main cause of farmer agitation in the district.

Malda District
Number of Farmer Attended Capacity Building
Programs Ever 2018

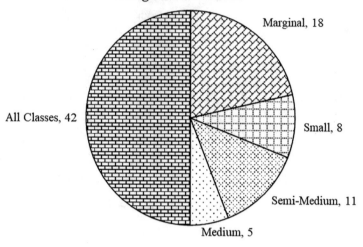

Fig. 5.39 Number of farmers who attended capacity building programs ever in Malda District 2018. (Source: based on field survey 2018 (see Annexures 5c–5g))

Malda District
Number of Farmer Ever been Participated to any
Agricultural Agitation 2018

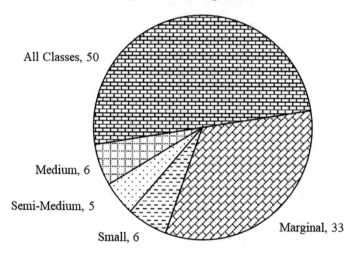

Fig. 5.40 Number of farmer ever been participated in agricultural agitation in Malda District 2018. (Source: based on field survey 2018 (see Annexures 5c–5g))

Malda District
Average Family Member of Households 2018

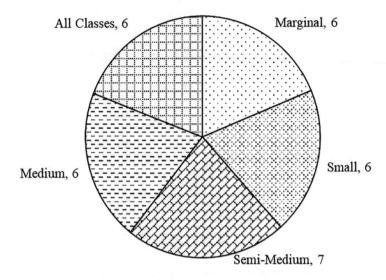

Fig. 5.41 Average size of households in Malda District 2018. (Source: based on field survey 2018 (see Annexures 5c–5g))

Average Size of Households
Except semi-medium land size class, average family size for other classes is six. The higher number of family labor helps in reducing the factor cost of production. But excess laborers create disguised unemployment in the agricultural system. In this sense, marginal farmers' families are having more disguised unemployed labor than the medium and semi-medium class (Fig. 5.41).

Average Age
The presence of the young age group (15–40 years) farmer leads agriculture more progressive, but due to migration, the male workforce left agriculture, and it makes the job of left behind aged farmers. For productive and innovative agricultural practices, the engagement and participation of young and adult age labor forces are essential. Figure 5.42 shows that the average age of sampled respondents under small (52 years) land size farmers is highest compared to the rest of the categories.

5.5   Agriculture Development and Crop Diversification in Sampled Villages

In Chap. 3, we have found the positive and significant relation between non-physical inputs of agriculture and crop diversification. The agricultural inputs like irrigation, fertilizers, machinery, and seeds show strong correlation with crop diversification

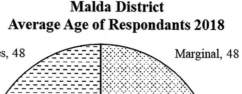

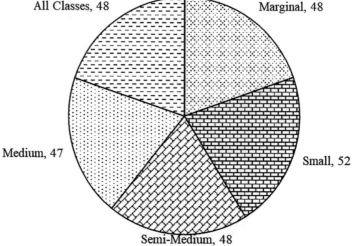

Fig. 5.42 Average age of respondents in Malda District 2018. (Source: based on field survey 2018 (see Annexures 5c–5g))

across land size categories in the district. In this section, an attempt has been made to look into the spatial relation of agriculture development with crop diversification in sampled villages.

The common pattern from land size class wise map ensures that the villages score above upper quartile bestowed with proper irrigation facilities, access to market, HYV seeds, mechanized farming, and socioeconomic quality of farmers. On the other hand, higher crop diversification is found in those villages where actual diversification is dominant under vegetable, oilseed, and fruit crop categories. Because of shortage of irrigation, land size, soil quality, price of agro-product, farm income, and socioeconomic conditions, the low or moderate development and diversification are profound in the district.

5.5.1   Agriculture Development and Crop Diversification in Marginal Category

Figure 5.43 illustrates the relations between agriculture development and crop diversification under marginal land size category in sampled villages in district. The agriculture development in six villages are having more than third quartile (> 0.527) score, four villages are below the lower quartile (< 0.467) range, and,five villages have been categorized under in medium CSS range (0.467–0.527).

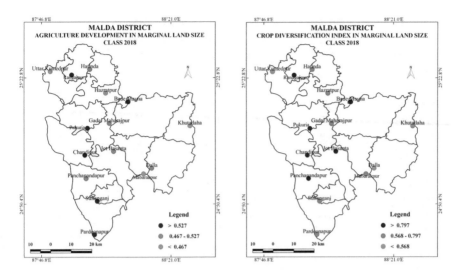

Fig. 5.43 Agriculture development and crop diversification under marginal land size class in Malda 2018. (Source: based on Tables 5.3 and 5.4)

Table 5.3 Composite standardized score of agricultural inputs in selected villages 2018

Villages	Marginal	Small	Semi-medium	Medium	All classes
Khutadaha	0.450	0.452	0.512	0.458	0.486
Hatinda	0.520	0.538	0.508	0.530	0.527
Hazaratpur	0.452	0.531	0.521	0.501	0.504
Jot Basanta	0.470	0.530	0.516	0.543	0.525
Bade Mayna	0.530	0.511	0.463	0.464	0.528
Dalla	0.492	0.497	0.465	0.534	0.519
Rangaipur	0.530	0.491	0.427	0.507	0.528
Uttar Kumedpur	0.453	0.466	0.528	0.480	0.503
Sultanganj	0.530	0.502	0.519	0.444	0.531
Panchanandapur	0.498	0.371	0.474	0.455	0.501
Par Deonapur	0.531	0.529	0.543	0.438	0.532
Mabarakpur	0.436	0.466	0.482	0.467	0.514
Chandipur	0.533	0.532	0.530	0.534	0.530
Gadai Maharajpur	0.513	0.465	0.544	0.470	0.528
Pukhuria	0.535	0.528	0.576	0.530	0.532

Source: Calculated by author (based on Annexures 5c–5g)
(Note: lower quartile = 0.467, upper quartile = 0.527)

In the district, the crop diversification above the upper quartile (> 0.797) index is found in six villages, low crop diversification noted in three villages (lower quartile range < 0.568), and six villages come under medium (0.568–0.797) quartile range. In four villages (Rangaipur, Bade Mayna, Pukhuria, and Chandipur), the higher agriculture development (score above the third quartile) boosts into higher crop

Table 5.4 Gibbs-Martins' crop diversification index in selected villages 2018

Villages	Marginal	Small	Semi-medium	Medium	All classes
Khutadaha	0.567	0.509	0.483	0.477	0.627
Hatinda	0.797	0.819	0.568	0.791	0.747
Hazaratpur	0.723	0.808	0.789	0.707	0.785
Jot Basanta	0.803	0.821	0.760	0.808	0.815
Bade Mayna	0.799	0.566	0.794	0.623	0.802
Dalla	0.482	0.447	0.452	0.364	0.546
Rangaipur	0.817	0.504	0.430	0.528	0.813
Uttar Kumedpur	0.540	0.801	0.418	0.701	0.615
Sultanganj	0.762	0.723	0.792	0.817	0.754
Panchanandapur	0.804	0.552	0.701	0.402	0.772
Par Deonapur	0.794	0.817	0.794	0.819	0.802
Mabarakpur	0.627	0.714	0.624	0.718	0.714
Chandipur	0.835	0.828	0.843	0.802	0.827
Gadai Maharajpur	0.729	0.782	0.797	0.836	0.796
Pukhuria	0.800	0.794	0.792	0.507	0.789

Source: Calculated by author
(Note: lower quartile = 0.568, upper quartile = 0.797)

diversification (above the third quartile). At the same time, the medium development and diversification are found in two villages mainly Gadai Maharajpur and Hatinda. The index value below the upper quartile in case of agriculture development and crop diversification shows in Khutadaha and Uttar Kumedpur in the district. Though, Panchanandapur is in medium range of agriculture development but it is reported into higher range of crop diversification. Par Deonapur village shows higher development but not higher crop diversification. Jot Basanta and Mabarakpur are such villages which do not fall on the above pattern, i.e., higher agriculture inputs led to higher crop diversifications.

5.5.2 Agriculture Development and Crop Diversification in Small Category

Agriculture development and crop diversification under small land size class category have been presented in Fig. 5.44. It shows that six villages lie above third quartile (>0.527), four in medium (0.467–0.527), and five villages under lower quartile range in agriculture development. The high crop diversification (above the third quartile index >0.797) is found in eight villages. Only two villages fall in medium range score (0.568–0.797), while five villages appear under first quartile index in the district.

The agriculture development and crop diversification are higher in (above third quartile) six villages, namely, Hatinda, Hazaratpur, Pukhuria, Chandipur, Jot

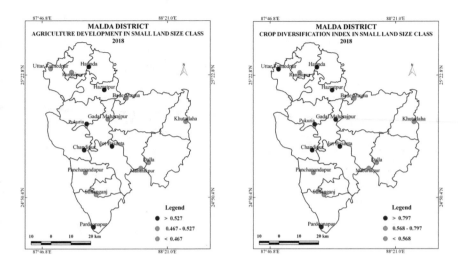

Fig. 5.44 Agriculture Development and crop diversification under small land size class in Malda District 2018. (Source: based on Tables 5.3 and 5.4)

Basanta, and Par Deonapur due to extended irrigation, HYV seeds, and fertilizers. Uttar Kumedpur village shows high diversification index but it lies below the first quartile in agriculture development. Here, the high crop diversification is mainly found because a variety of crops grow under vegetables. Sultanganj and Mabarakpur fall in the same category in case of development and diversification in the district. Panchanandapur and Khutadaha have reported low development and low diversification due to cereal crop dominating gross cropped area. The medium development is rapt to low diversification in case of Dalla village, and medium development leads to low diversification for Rangaipur and Bade Mayna in the district.

5.5.3 Agriculture Development and Crop Diversification in Semi-Medium Category

With five villages in agriculture development and four villages in crop diversification, the higher development and diversification under semi-medium land size class category have been presented in Fig. 5.45. The agriculture development with medium range score reflects in seven villages and crop diversification index in six. Three villages score below first quartile in agriculture development, while five villages come under 0.568 index value in crop diversification. The higher agriculture development led to higher crop diversification in Gadai Maharajpur, Chandipur, and Par Deonapur in the district. Hazaratpur, Jot Basanta, Sultanganj, Panchanandapur, and Mabarakpur show medium development and crop diversification. Rangaipur

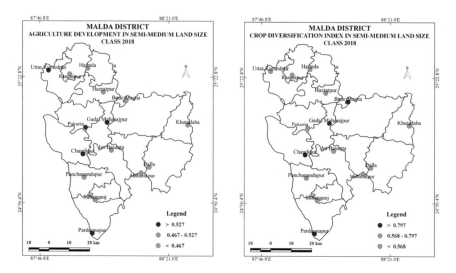

Fig. 5.45 Agriculture development and crop diversification under semi-medium land size class in Malda District 2018. (Source: based on Tables 5.3 and 5.4)

and Dalla are such villages where low agriculture development led to low crop diversification. Bade Mayna falls in low agriculture development but ensures high crop diversification due to large numbers of crops under oilseeds and pulses. Khutadaha village has medium development but low crop diversification. The villages that lie above upper and medium quartile range have also ensured better institutional and infrastructural facilities and higher crop diversification.

5.5.4 Agriculture Development and Crop Diversification in Medium Category

Under medium land size category, the higher development and diversification have been found in four and five villages, respectively. The medium range development and diversification report in five and four villages accordingly in the district. The lower quartile index has been observed in six villages in agriculture development, and four villages fall in low crop diversification (Fig. 5.46) category. Jot Basanta is the only village in the district where high development and diversification are identical due to higher diversification under fruits and vegetables. Hatinda and Hazaratpur are such villages in which medium range development has led to medium diversification index. Khutadaha and Panchanandapur have lower development and lower crop diversification due to shortage of irrigation. The transformation is found in Dalla, Pukhuria, and Hatinda where high and medium development led to low and medium crop diversification. Sultanganj and Par Deonapur fall in low development and higher crop diversification due to dominated diversification under vegetables.

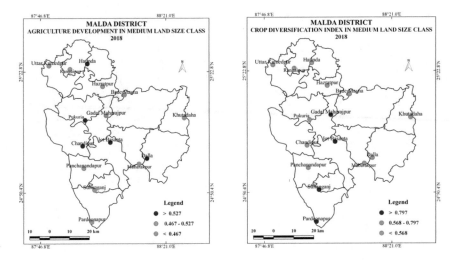

Fig. 5.46 Agriculture development and crop diversification under medium land size class in Malda District 2018. (Source: based on Tables 5.3 and 5.4)

5.5.5   Agriculture Development and Crop Diversification in All Land Size Category

Under all land size classes, the higher agriculture development (six villages) and crop diversification (seven villages) have been found in six and seven villages, respectively. The medium range development and diversification reflect in six and five villages (Fig. 5.47). The low agricultural development and crop diversification report in three villages. The higher agriculture development encourages higher crop diversification which is found in five villages, namely, Rangaipur, Gadai Maharajpur, Pukhuria, Chandipur, and Sultanganj in the district. The medium range development and diversification shows in four villages like Uttar Kumedpur, Hatinda, Bade Mayna, and Panchanandapur. And, the lower agriculture development can be correlated with lower crop diversification in cases of two villages like Dalla and Khutadaha. The high agriculture development induces low crop diversification in Jot Basanta, while higher development leads to medium diversification in Sultanganj village. In Mabarakpur village, medium development leads to low diversification in the district.

The overall analysis on agriculture development and crop diversification though found a mixed pattern, but, in most cases, it is rightly correlated with higher development and higher diversification. Across different land size classes, high agriculture development (> 0.527 CSS) is recognized because of endowed irrigation facilities, HYV seeds, mechanized farming, fertilizers, and socioeconomic aspects have boosted crop diversification under different crop categories. In sampled villages crop diversification is maximum under vegetable crops which contribute positively in total diversification index.

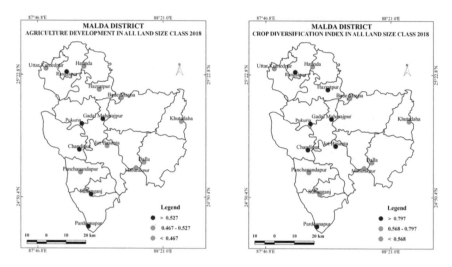

Fig. 5.47 Agriculture development and crop diversification under all land size class in Malda District 2018. (Source: based on Tables 5.3 and 5.4)

5.6   Analysis of Interdependence

In geographical studies, most systems are described by a large number of variables, and most of these variables are interrelated. Hence, information has duplicated that mark the essential structure of the system. Interdependence of the variables is analyzed via two approaches: principal component analysis (PCA) and factor analysis (FA). PCA offers a means of removing redundancies from a set of consistent variables, and the resulting principal components are uncorrelated. FA, on the other hand, is a method of investigating the correlated structure of a multivariate system. Thus, it is an attempt to find groups of the variable (factors) measuring a single important aspect of the system. As these factors are not necessarily uncorrelated, a method of transforming these factors called rotation is applied.

5.6.1   Principal Components Analysis (PCA)

The PCA is one of the factor reduction techniques which is used to construct a new variable from uncorrelated pairwise variables under operation. In the first step, the strongly correlated variables are determined and then remaining less correlated variables. The remaining latent (less) variables are termed as factor or component; therefore it is known as factor analysis (Helmy and El-Taweel 2009). This was first initiated in 1901 by Perason and as a model developed by Hotelling (1933) with the objective to replace large data sets into smaller datasets.

In present study, PCA calculation is done in Jamovi 0.9.2.6 software. The mathematical equation in the present model follows linear functions which are later chosen for orthogonal analysis. The final principle components are composite of original variables in which first component has maximum variance and second component is the orthogonal to first component which derived from maximum variance of all linear functions in given variables. In the present model, varimax rotation technique is used to get rotated matrix for better interpretation of factors to opt for reasonable conclusions. The equation applied here is

$$F* = ZAD^{-1/2}$$

where F* replaced by F which is the standardized unit variance. Here, Z represents standardized equivalents of X variables. A is the matrix of unit-length eigenvectors R which derives from the matrix (X'X) and becomes R as a correlation matrix. And, D drives from a diagonal matrix of eigenvalue of X'X.

In present analyses, 32 variables have been categorized under 8 indicators which are further reduced into components. As this noted, agriculture development and crop diversification are significantly correlated. It means variables or factors which influence crop diversification have also impact on agriculture development. Since the number of variables is 32, therefore, these are further classified and reduced under component through principal component analysis.

The first three components explain 78.5% in total variance under eight selected indicators in marginal land size class category (Table 5.5). In component loading, the first component explains 35.2% of total variance, and it is 24.7% under the second component, while 18.6% is explained under the third component. Under the first component, market and socioeconomic loaded positively, while in second component, income and credit and machinery have loaded positively. In the third component, soil and fertilizers found maximum loading in marginal land size class category.

In the small land size class category, the first three components explain 68.8% of total variance. Under the first component, socioeconomic and cold storage are loaded positively which fetches the sum of the square of the component up to 2.18 and explains 27.3% of total variance (Table 5.5). The components like seeds, income, credit, and socioeconomic are positively loaded. The second component explains 24.9% with 1.99 sum of square. With 1.32 sum of the square of the component, socioeconomic is loaded positively, and the third component explains 16.6% in total cumulative variance under small land size class.

Under three components, the semi-medium land size class category explains 61.2% of cumulative variance. Under first component soil and fertilizers, market and storage and socioeconomic are loaded positively and cumulatively add 2.18 as a sum of the square of the first component. In the same category, the second component is loaded 1.39 as a sum of the square of the component in which seeds, income, and credit and socioeconomic have loaded positively which pays 17.4% in total cumulative variance. The third component shares 16.6% in cumulative variance where the sum of a square of components contributes 1.39 and the indicators like socioeconomic are loaded positively in it (Table 5.5).

Table 5.5 Principal component loading (PCA)

Indicators	Marginal component			Small component			Semi-medium component			Medium component			All classes component		
	1	2	3	1	2	3	1	2	3	1	2	3	1	2	3
Land	−0.563	−0.387	−0.497	−0.848			−0.848					0.838	−0.617	−0.335	0.579
Irrigation	0.399	−0.468	0.397		−0.322			−0.322			0.631	−0.373	−0.742		
Seeds	−0.792		0.428		0.613			0.613		0.513		0.667			0.686
Income and credits	0.374	0.791			0.77			0.776		0.837			0.508	0.344	
Machineries		0.763		−0.681			−0.681				−0.893		0.522	−0.573	−0.432
Soil and fertilizers			0.911	0.313	0.911	−0.841	0.313		−0.841	−0.505	0.558	0.318		0.916	
Market and storage	0.889			0.792			0.792			−0.805			−0.307		−0.674
Socioeconomic	0.876			0.325	0.378	0.719	0.325	0.378	0.719	0.678	0.517		0.878		
SS loading	**2.82**	**1.98**	**1.49**	**2.18**	**1.99**	**1.32**	**2.18**	**1.39**	**1.32**	**2.41**	**1.87**	**1.47**	**2.35**	**2.01**	**1.46**
% of variance	**35.2**	**24.7**	**18.6**	**27.3**	**24.9**	**16.6**	**27.3**	**17.4**	**16.6**	**30.1**	**23.4**	**18.4**	**29.4**	**25.2**	**18.2**
Cumulative %	**35.2**	**59.9**	**78.5**	**27.3**	**52.2**	**68.8**	**27.3**	**44.6**	**61.2**	**30.1**	**53.5**	**71.9**	**29.4**	**54.6**	**72.8**

Source: Computed by author

Note. "Varimax" rotation was used

The medium land size category explains 71.9% total cumulative variance under three components. With 2.41 as a sum of a square of components is loaded under first components, income and credits, seeds, and socioeconomic are loaded positively and explain 30.1% of total cumulative variance (Table 5.5). The second component explains 23.4% in total variance with 1.87 as a sum of a square of component contributes positively in irrigation, socioeconomic, soil, and fertilizers. The third component in the medium land size category explains 18.4% of total cumulative variance with positive loading of land, seeds, soil, and fertilizers which fetch 1.47 as a sum of squares.

Under all land size classes, three components explain 72.8% cumulative variance. With positive loading from machinery, income and credits, and socioeconomic, the first component explains 29.4% of the total cumulative variance and 2.35 as the sum of a square of components. The second component accounts positive loading of soils, fertilizers, and machineries which explains 25.2% of total cumulative variance (Table 5.5). The sum of the square of the component from second and third component loading is 2.01 and 1.46, respectively. With positive loading of land and seeds, the third component explains 18.2% in total cumulative variance.

The cumulative variance suggests that the agriculture development and crop diversification are best explained under marginal and medium land size categories. The PCA for both also indicates that agriculture development in overall is not but some selected variables are exclusive for crop diversification under small and semi-medium land size categories.

5.7 Measuring Agriculture Efficiency

The word efficiency simply means input-output ratio which is synonymously used to refer to productivity. Input-output performance of agriculture indicates the capacity of a farm in terms of production ability. For example, in a given input if output is low as compared to other regions or farms, then input generally increases to get desired output. In any production system, inputs generally combine with more than one set of variables, and an increase in single input may impact overall output. Therefore, an output is an outcome from many inputs or sometimes more than one output drawn from a single input. In a such case, efficiency measurement is an absolute step where all inputs combine into one to total output. Technically, this input-output mechanism has stark limitations especially when single input oriented more than one output and if different input results in different amounts of output and then performance of different farms or regions with respect to one another is near to impossible to measure. Here, to check the relative performance of different farm units, data envelope analysis (DEA) is used.

DEA is the nonparametric and linear programming method used to measure productive performance of decision-making units (DMU). The DEA differs from simple efficiency in that multiple inputs and output accommodates under analyses instead just input-output ratio. Productive performance of different DMU is checked through technical efficiency scores between 0 and 1. The technical efficiency rating 1 is efficient and assumed as best practice where efficiency rating less than 1 evaluated as inefficient. In present study, the DEA runs in DEA solver. The DEA model is

$$\max \theta_k = \sum_{r=1}^{s} w_r y_{rk.}$$

subject to

$$\sum_{i=1}^{m} \mu_i x_{ik} = 1$$

$$\sum_{r=1}^{s} w_r y_{rj} - \sum_{i=1}^{m} \mu_i x_{ij} \leq 0 \quad j = 1,\ldots,N$$

$$w_r \geq \varepsilon, \ \mu_r \geq \varepsilon \quad r = 1,\ldots,s \ \ i = 1,\ldots,m$$

In the model, total set of N units in which m inputs and s outputs. In unit j, i input denotes x_{ij} and output as y_{rj}. This model is a multiplier form where output weights w_r and input as μ_r. The model seeks to optimize maximum and minimum efficiency $\theta \leq 0$ *and* = 1 of j units. In the model, input and output weight to be positive and greater than value ε.

5.7.1   *Relative Efficiency Under Different Land Size Category*

In the study, DEAP computer program (1997) is used to calculate the technical efficiency of an output-oriented model which involves one input and three outputs. To calculate relative efficiency of agriculture at village level here, composite standardized score (CSS) of 32 variables have been taken as inputs; and Yang's yield index, crop diversification index and cropping intensity undertake as output. The data on yield index and cropping intensity attached in the annexure.

Table 5.6 reports that the result of DEA is applied on 15 decision-making unit (DMU) and these are villages in the present case. The DEA recognizes the same inefficiency village that is recognizable through observation of the data. Except Hazaratpur, Jot Basanta, and Rangaipur, all other villages have efficiency ratings below 1.00; hence except these villages, other DMU has been identified as inefficient.

The reason for variations in efficiency score among the village can be correlated with nature and quality of soils, cultivation techniques, cropping pattern, nature of crops grown, and quality of farmers. However, there is no significant variation in the use of amount inputs with respect to the size of the villages though there are variations in output.

Table 5.6 reveals that on the same amount of inputs, Khutadaha village has achieved relatively lower output with respect to Hazaratpur, Jot Basanta, and

Table 5.6 Relative efficiency score of marginal land size class 2018

Input	Output	
CSS	Yield index	
	Crop diversification index	
	Cropping intensity	

DEA-CRS model

Input-oriented

DMU No.	DMU name efficiency	Technical efficiency
1	Khutadaha	0.845
2	Hatinda	0.964
3	Hazaratpur	1.000
4	Jot Basanta	1.000
5	Bade Mayna	0.921
6	Dalla	0.798
7	Rangaipur	1.000
8	Uttar Kumedpur	0.995
9	Sultanganj	0.892
10	Panchanandapur	0.932
11	Par Deonapur	0.929
12	Mabarakpur	0.902
13	Chandipur	0.962
14	Gadai Maharajpur	0.911
15	Pukhuria	0.859

Source: Computed by author in DEAP computer program

Rangaipur villages. To be an efficient village (Khutadaha) in the present case, the same village has to reduce 1–0.845 = 0.155 or 15.50% input with constant present outputs (crop diversification, productivity, and cropping intensity). On same amount inputs, the efficient villages (such as Hazaratpur) have achieved comparatively better output. In the present analysis, Dalla is the most inefficient village in the marginal land category as it uses 0.202 or 20.02% (1–0.798) more input on the same output as compared to Jot Basanta and other efficient villages.

It appears in Tables 5.7 and 5.8 that in small land size class, two villages, namely, Jot Basanta and Panchanandapur, have been found as efficient villages (scored 1 or 100%) and Hatinda (0.727) arise as the most inefficient village in the district. Three villages, namely, Rangaipur, Panchanandapur, and Mabarakpur, are found as efficient, while Pukhuria with efficiency score 0.781 is found as the most inefficient village in semi-medium land size category in the district. In medium land size category, Sultanganj and Par Deonapur arise as efficient villages, and Jot Basanta (0.826) is the most inefficient village in the district. Under all land size classes, Rangaipur, Uttar Kumedpur, Panchanandapur, and Par Deonapur have been the efficient villages, while Dalla with 0.812 efficiency score arises as the most inefficient village in the district. In the district, Hatinda is the most inefficient village.

Based on Table 5.7, the final efficiency table has been prepared and presented in 5.8 Table to show total efficient and inefficient villages in the district.

Table 5.7 Relative efficiency score of different land classes 2018

Villages	Marginal	Small	Semi-medium	Medium	All class
Khutadaha	0.845	0.788	0.889	0.951	0.850
Hatinda	0.964	0.727	0.806	0.895	0.958
Hazaratpur	1.000	0.784	0.952	0.829	0.968
Jot Basanta	1.000	1.000	0.928	0.826	0.963
Bade Mayna	0.921	0.843	0.977	0.922	0.960
Dalla	0.798	0.789	0.873	0.883	0.812
Rangaipur	1.000	0.898	1.000	0.937	1.000
Uttar Kumedpur	0.995	0.782	0.923	0.926	1.000
Sultanganj	0.892	0.807	0.890	1.000	0.923
Panchanandapur	0.932	1.000	1.000	0.900	1.000
Par Deonapur	0.929	0.792	0.847	1.000	1.000
Mabarakpur	0.902	0.917	1.000	0.968	0.969
Chandipur	0.962	0.935	0.982	0.950	0.906
Gadai Maharajpur	0.911	0.890	0.828	0.939	0.901
Pukhuria	0.859	0.777	0.781	0.891	0.882

Source: Computed by author

Table 5.8 Final efficiency table 2018

Village	Marginal	Small	Semi-medium	Medium	All class
Khutadaha	Inefficient	Inefficient	Inefficient	Inefficient	Inefficient
Hatinda	Inefficient	Inefficient	Inefficient	Inefficient	Inefficient
Hazaratpur	**Efficient**	Inefficient	Inefficient	Inefficient	Inefficient
Jot Basanta	**Efficient**	**Efficient**	Inefficient	Inefficient	Inefficient
Bade Mayna	Inefficient	Inefficient	Inefficient	Inefficient	Inefficient
Dalla	Inefficient	Inefficient	Inefficient	Inefficient	Inefficient
Rangaipur	**Efficient**	Inefficient	**Efficient**	Inefficient	**Efficient**
Uttar Kumedpur	Inefficient	Inefficient	Inefficient	Inefficient	**Efficient**
Sultanganj	Inefficient	Inefficient	Inefficient	**Efficient**	Inefficient
Panchanandapur	Inefficient	**Efficient**	**Efficient**	Inefficient	**Efficient**
Par Deonapur	Inefficient	Inefficient	Inefficient	**Efficient**	Inefficient
Mabarakpur	Inefficient	Inefficient	**Efficient**	Inefficient	**Efficient**
Chandipur	Inefficient	Inefficient	Inefficient	Inefficient	Inefficient
Gadai Maharajpur	Inefficient	Inefficient	Inefficient	Inefficient	Inefficient
Pukhuria	Inefficient	Inefficient	Inefficient	Inefficient	Inefficient

Source: Based on Table 5.7

5.7.2 Efficiency Determination

It is proven fact that agriculture inputs increased agricultural output up to certain extent, and it is for sure in developing agricultural systems. But, can higher agricultural inputs lead to efficient agriculture? Except Rangaipur in marginal and all land

Table 5.9 Efficiency determinates test: inputs and outputs

t-test: two-sample assuming unequal variances		
	CSS	*Efficient*
Mean	0.552564	1
Variance	0.001593	0
Observations	75	15
Df	74	
t stat	−97.0903	
P(T < =t) two-tail	0.00000	
t critical two-tail	2.643913	

Source: Computed by author

size class and Jot Basanta in small land size category, all other villages show higher agriculture inputs (above third quartile) but spot as an inefficient village (Table 5.8). In present study, the underlying assumption is that the higher agriculture inputs make agriculture efficient which has been tested statistically in two tailed t-test (in Excel) and presented in Table 5.9.

Table 5.9 reports assumption check summary based on t-test which shows a significant difference in the mean score of agricultural inputs and efficiency score at high critical table value, i.e., 2.643913. Therefore, the underlying assumption fails to accept the proposed hypothesis.

References

Hall, T., & Healey, M. (2005). Disabled students' experiences of fieldwork. *Area, 37*(4), 446–449.

Helmy, A. K., & El-Taweel, G. S. (2009). Authentication scheme based on principal component analysis for satellite images. *International Journal of Signal Processing, Image Processing and Pattern Recognition, 2*(3), 1–10.

Siddiqui, S. H. (2007). *Fifty years of Indian agriculture* (p. 176). New Delhi: Concept Publishing Company.

Chapter 6
Challenges and Opportunities of Agricultural Development in the District

6.1 SWOT Analysis

The SWOT (strengths, weaknesses, opportunities and threats) analysis is a device that helps in framework development to researchers, planners, policy analysts, academic think-tanks, and business managers to identify, to formulate, and to prioritize the strategies for the possible solutions. According to the Agriculture Census of 2015–2016, Malda district of West Bengal is the home of more than 96 per cent marginal and small size (land) farmers. The intensive low subsistence cum commercial agriculture is the only boost for income and livelihood of rural folk. In this context, SWOT analysis is used to identify the strategy for agricultural development. The identification of strengths and opportunities is necessary to reduce the risk of internal weaknesses and the external threats of agriculture in the district. In order to make the chart of a course for redevelopment of agricultural sector and its higher contribution in district economy, the analysis will help in developing a strategic plan to make the agricultural sector remain competitive.

The SWOT analysis is based on the consideration of factors from social, economic, and environmental aspects especially which have direct and indirect link to agriculture. The analysis will provide necessary inputs toward the formulation of district plan for agricultural development in general and holistic development in particular. Information on SWOT at block and district is catered out from Agriculture Development Officer (ADO) and Deputy Director of Agriculture (Admin.), respectively. The block level SWOT information has been obtained in February 2018. The SWOT questionnaires were open-ended in nature. The different types of strengths, weaknesses, opportunities, and threats have been listed in Table 6.1.

Table 6.1 SWOT analysis for agriculture in Malda District 2018

Strengths	Weakness
Cropping intensity increased	Very limited agro-based industries
Micro ecological opportunities	Very low institutional credit
Abundant sources of freshwater	Soil health degrading
Agro based small industries	Low level of farm mechanizations
Crop under high-value crop increasing	High cost of cultivation
Diversification increasing	Limited infrastructure and marketing
Means and medium transport improving	Havoc run-off of monsoon rain
Opportunities for diverse outsourcing	Age-old cultivation methods
Opportunities	**Threats**
Agro-based industries	Erratic rainfall during Kharif season
Organic farming	Flood and soil erosion
The scope of introduction of more HYV seeds	Increasing night time temperature
Involvement of SHGs, NGOs, private org.	Overdoses of chemical fertilizers
Promotion of demand driven production	Hailstorms and Kalbaisakhi
Introduction of non-conventional crops	Sudden outbreak of crop's diseases.
Intensification of high-value crops	Lowering water Table.
Soil test value-based nutrition management	Sudden price fall of agro products

Source: field survey 2018

6.1.1 Strengths

The district has a lot of aquatic potential sources such as rivers, ponds, beels, and wetlands which provide adequate irrigation for diverse agriculture. Apart from that the district has an advantage of alluvial soil, tropical monsoon climate, and huge population dividend (15–40 years age-groups). However, from the field survey, the following specific strengths have been identified:

1. The strength of agriculture in Malda has been recognized in its diverse agro-climatic region extended from north to south.
2. The cropping intensity of the district has increased almost twofold from 1990 to 2016. Productivity and yield of cereals have raised 2.5-fold, while other crops like oilseeds, fruits and spices report an increase in area and production.
3. Malda district receives monsoon rainfall every year and enormous minor rivers are enriched with freshwater. The monsoon rainfall also boosts pisciculture and rain-based farming.
4. The agro-based industries especially cereal crops-based industries are increased, but it is only localized improvement.
5. The areas under high value crops, such as makhana cultivation in wetlands of Tal region, cocoon farming for sericulture and some medicinal crops in Diara region, and floriculture in Old Malda and Gazole, have replaced traditional paddy crops. As a result, the income of the farmers has increased many folds.

6. DDA (Admin.) states that agriculture and crop diversification have increased in the district. Along with the crop cultivation, farmers of Diara region (mostly) also practice animal husbandry and milk mulching. The number of crops has increased almost three folds from 1995–1996 to 2015–2016 which indicates there is increasing crop diversification in the district.

7. The means and medium of transport such as road transportation network and railways have been improving continuously from 1990 which fetches cheap and efficient transportation that ultimately benefits in purchase and sale of agricultural commodities.

8. Apart from these, the labor work forces' emigration has opened up a new avenue of outsourcing to marginal and small land size farmers. The opening of small and medium shops boosts opportunities for employment. The technological innovations have brought a new source of outsourcing to farmer households which eventually augments agriculture in the district.

6.1.2 Weaknesses

Despite the enormous strengths inherent in the agriculture system of Malda, its efficacy in contributing to overall development is fraught in some factors:

1. Malda district is known for world-famous mango but recent trends suggest the declining trends of demand in national and international markets due to losing quality. The lack of mango allied industries, it retails mostly in the local markets which is not profitable for the mango growers. The vegetable, oilseed, fruit, fiber, and cereal crop-based industries are almost lacking in the district; therefore secondary and tertiary activity-based agricultures have yet not groomed so far.

2. The agricultural credit is decreasing as it guarantees only to medium and semi-medium land size farmers. The Regional Rural Bank (RRB) has been reduced to a credit amount which was earlier disbursed to the farmers for the sake of agricultural innovations.

3. The heavy dose of insecticides, pesticides, and fertilizers makes agriculture intense which ultimately effects on soil health negatively reported by DDA (Admin.), Malda.

4. The low socioeconomic condition and small parcel of land have caused a hindrance in adoption of mechanized farming.

5. The increasing cost of production (inputs) have slowed agricultural outputs.

6. The market infrastructure, especially regulated markets, fair price shops, public godowns, and cold stores, is very few in number and limited in one to two blocks only. And that is why farmers do not get the benefit of minimum support price (MSP) fixed by the government every year.

7. The district receives sufficient rainfall but there is a shortage of irrigation from February to April especially for vegetables and pulse crops in the Barind region of the district.

8. The methods of cultivation are also age-old and the adaptation of innovations is very slow.

6.1.3 Opportunities

There are abundant opportunities for the agriculture in Malda to bounce back to a place of reckoning with the sector:

1. The opportunities include establishments of agro-based industries which could easily be worked back into rejuvenating the agricultural sector.
2. The district has enough potential to feed the growing demand of organic vegetables at local and national markets. Organic farming is also an opportunity to maintain soil quality which ultimately fetches low-cost farming.
3. Although HYV seeds have been used by the farmers to a broader extent, still its use could be extended to some crops like pulses, fruits, oilseeds, and fibers in the district.
4. Involvement of the self-help group (SHG), non-profitable organizations, NGOs, and private extension service groups could be linked with the performance of poor farmers and their farming system. For that, farmer club and community extension service would be a viable option.
5. The promotion and production of high-value crops (fruits, vegetables, makhana, sericulture, oilseeds, and pulses) can be replaced over traditional crops.
6. The introduction of non-conventional crops such as drugs and narcotic crops could be a feasible option for a better upshot to the farmers.
7. There are ample opportunities to extend and intensify high-value crop (spices, pulses, oilseeds, flowers, and makhana) areas in some blocks.
8. To minimize the risk of unscientific cultivation, soil test-based nutrition management is an essential future option for soil health.

6.1.4 Threats

A threat, in the context agriculture, refers to anything that has the potential to cause serious harm to the agricultural system. A threat is something that may or may not happen but has the potential to cause serious damage. The recurrent threats facing by the agriculture in Malda are noted below:

1. The Erratic rainfall distribution during kharif crop season results in huge crop loss and cropping pattern in the district.
2. Almost every year, flood, drought, and soil erosion are the major problems in nine blocks of the district.
3. From the last 30 years, there has been an increasing trend in night time temperature. It hampers plant's growth in their life course. Apart from that, increase in temperature especially terminal heat during rabi season is another concern.
4. The overdose of chemical fertilizers especially to marginal and small land size farmers is a key concern for sustainability of soil health.

5. Every year during the month of April and May, the hail storms and Kalbaisakhi have damaged existing cropping patterns, resulting in huge crop loss especially fruits (mango, litchi), cereals (paddy), vegetables, spices and fibers (jute) crops. As the intensity of hailstorms fluctuates every year, therefore, the nature of losses varies.
6. Very often, there are some of the crops affected by diseases. The frequent and recurrent attack of certain insects and pests in cereals, vegetables, and fruits is unprecedented in the last few years which results in massive crop loss. The effect of crop diseases might likely be more pronounced as a result of unscientific methods of cultivation – reported by DDA (admin.) Malda.
7. For the sake of intensive subsistence agriculture, the groundwater extraction is increasing day by day. The decreasing groundwater table especially in Barind region of the district is very alarming.
8. Farmers are often in fears of instability and volatility of agro-products price as they have experienced many times in the past.

6.2 SWOT Mapping

An attempt has been made to list out the number of strengths, weaknesses, opportunities, and threats based on extensive discussion with agriculture development officers as well as different land size farmers in the district. Since the SWOT questionnaires were open ended in nature, therefore, a number of strengths (24), weaknesses (23), opportunities (24), and threats (20) are not equal under each particular. Each item of SWOT has been listed in alphabetical order in Table 6.2.

Arrangement of the SWOT for planning purposes has further been classified into internal (strengths and weaknesses – Table 6.3) and external (opportunities and threats – Table 6.4). To identify the extent which builds on strengths and minimizes the weakness is a critical area that involves the strategies in relation to its competitor (here other agricultural systems). The opportunities and threats are influenced by the external environment, and, therefore, focusing on the proper evolution of strategic intention is totally dependable to others aspects.

6.2.1 Strengths Mapping in Malda District

Some of the numbers that appear very frequently in Table 6.3 and Fig. 6.1 are access of financial institutions in almost every block in the district which indicates the chance of financial transactions along with agriculture loans and credit to the farmers. In five blocks, the number one appears as the best strength, namely, in Ratua-I, Gazole, Habibpur, Old Malda, and Ratua-II in the district. The available groundwater stood as number one strength in six blocks like Harishchandrapur-II, Chanchal-I, II, Manikchak, Kaliachak-II, and Kaliachak-III in the district.

Table 6.2 Ranking of SWOT in agriculture in Malda District 2018

Sl.	Strengths	Sl.	Weaknesses	Sl.	Opportunities	Sl.	Threats
1	Access to financial institution	1	Child labor	1	Agricultural diversification	1	Crop diseases
2	Availability of groundwater	2	Deteriorating soil quality	2	Agro-based industries	2	Decreasing wetlands
3	Availability of organic waste	3	Heavy doses of pesticides	3	Crop diversification	3	Water shortage
4	Conservation of rainwater	4	The high cost of cultivation	4	Crop variety on rotation	4	Erratic rainfall
5	Crop productivity increased	5	Lack of new technology	5	Cultivation of high-value crops	5	Excessive use of fertilizers
6	Cropping intensity increasing	6	Lack of industries	6	Demand based production	6	Flood
7	Demand driven farming	7	Lack of goods and cold storage	7	Dry crops cultivation	7	Fluctuation in price
8	Fertile soil	8	Lack of interest in new farming	8	Efficient irrigation systems	8	Increasing emigration
9	Fodder cultivation	9	Lack of regulated market	9	Export oriented cultivation	9	Increasing temperature
10	Good number dealers	10	Land fragmentation	10	Extension of fodder cultivation	10	Inorganic farming
11	A good number of cattle	11	Low returned	11	Extension of jute factory	11	Kalbaisakhi and hailstorms
12	Good road network	12	Low yield	12	Fallow land to be cultivated	12	Land encroachment
13	High-value crops farming	13	Lowering water tables	13	Fare and regulated market system	13	Land security
14	Mango based industry	14	Marginal poor farmers	14	Field crop processing units	14	Rainfall based farming
15	Outsourcing increasing	15	Overdoses of fertilizer	15	Increase productivity	15	Soil erosion
16	Plenty of labor	16	Poor farmers	16	Increased extension service	16	Sudden fall of price
17	Presence of market	17	Poor road and transport	17	Intensive and mixed cultivation	17	Unexpected rainfall
18	Presence of water bodies	18	Rain-fed agriculture	18	Involvements of NGOs	18	Unexpected rain
19	Self-help group (SHG)	19	Reducing government incentives	19	Nearness to urban market	19	Water scarcity
20	Silt deposition on land	20	Slow adaptation of new change	20	Organic farming	20	Waterlogged
21	Small agro-Industry	21	Traditional crop farming	21	Plantation based Agriculture		
22	Use of organic manure	22	Unscientific farming	22	Road and railway connectivity		
23	Water retention is high	23	Workforce migration	23	Soil test nutrition management		
24	Women participation			24	Wetland-based cultivation		

Source: field survey 2018

Table 6.3 Arrangement of strengths and weaknesses (internal) in agriculture in Malda District 2018

Block	Strengths								Weaknesses							
Bamongola	2	3	5	6	18	20	23	24	4	5	9	11	13	16	21	22
Chanchal-I	2	3	9	12	14	16	17	18	4	7	8	10	15	17	18	20
Chanchal-II	2	3	5	8	16	17	18	23	3	4	6	8	11	14	16	19
English Bazar	4	5	6	11	16	19	22	24	3	6	11	13	15	19	21	23
Gazole	1	2	5	8	13	16	20	24	2	4	6	13	14	19	20	21
Habibpur	1	3	5	6	12	13	17	21	2	12	13	16	18	19	21	23
Harishchandrapur-I	5	8	10	16	18	19	20	23	8	10	14	15	16	19	20	21
Harishchandrapur-II	2	5	6	10	14	16	18	23	3	4	6	8	12	16	20	22
Kaliachak-I	5	10	11	12	13	16	19	24	2	3	10	11	13	15	19	20
Kaliachak-II	2	5	8	10	15	18	20	23	4	6	9	11	14	17	18	21
Kaliachak-III	2	12	14	17	19	21	22	24	1	3	5	7	9	15	22	23
Manikchak	4	8	8	9	11	15	19	24	6	9	10	12	15	17	19	23
Old Malda	1	4	5	6	9	10	12	17	1	2	7	13	16	18	19	22
Ratua-I	1	5	7	9	11	16	19	24	4	6	7	10	11	14	15	18
Ratua-II	1	2	5	10	14	16	20	24	4	6	8	9	15	18	20	22

Source: field survey 2018 (see Table 6.2)

Table 6.4 Arrangement of opportunities and threats (external) in agriculture of Malda District 2018

Block	Opportunities								Threats							
Bamongola	5	7	8	9	13	15	17	20	1	3	5	9	11	13	14	19
Chanchal-I	1	2	5	6	10	11	13	20	1	2	4	6	9	11	16	17
Chanchal-II	1	2	3	4	5	15	17	23	5	6	9	10	11	16	17	20
English Bazar	3	5	6	8	14	19	20	24	1	2	3	4	4	11	16	17
Gazole	3	7	8	10	12	16	17	21	1	3	5	7	9	10	11	14
Habibpur	2	7	8	9	10	15	17	20	4	7	8	9	10	11	12	17
Harishchandrapur-I	3	4	9	13	17	20	23	24	3	5	6	7	9	11	16	20
Harishchandrapur-II	1	2	3	13	17	20	23	24	2	4	6	7	9	11	16	20
Kaliachak-I	1	6	7	8	18	19	20	23	1	2	3	4	5	11	12	17
Kaliachak-II	2	3	6	9	15	17	18	22	1	3	4	9	11	16	16	17
Kaliachak-III	2	3	10	17	18	21	23	24	1	2	3	3	6	11	15	17
Manikchak	2	3	4	10	13	17	18	21	1	2	4	5	10	11	16	17
Old Malda	2	3	4	6	7	8	15	19	1	3	4	9	11	12	16	19
Ratua-I	2	3	4	6	9	15	17	20	1	3	4	9	10	11	16	17
Ratua-II	3	5	10	13	17	18	20	21	1	1	3	7	9	11	14	16

Source: field survey 2018 (see Table 6.2)

The water conservation for irrigation and drinking purposes is the number one strength in some of the blocks like English Bazar and Bamangola in the district. Available organic wastes (cow dung, ashes, and others) locate as a strength in four blocks, but in Bamangola, Habibpur, Chanchal-I, and Chanchal-II block, it placed as number second strength. Crop productivity substantially increased across the blocks in the district, and it's mainly because of the agricultural innovation.

Cropping intensity is one of the important aspects depicted as a strength in six blocks which is reported by Agriculture Development Officers of respective blocks.

The demand-driven agricultural practice is infrequent as a strength although it shows in Ratua-I block under the spice crops. Fertile soil is one of the important strengths in five blocks. The fodder cultivation shows as a strength in Manikchak, Ratua-I, and Chanchal-I block in the district.

A good number of agro-dealers whether it is a fertilizer or purchase dealer have been found as strengths in Kaliachak-I, Kaliachak-II, Ratua-II, Harishchandrapur-I, and Harishchandrapur-II block. A good number of cattle including cows and buffalos have dominance in share in blocks like English Bazar, Kaliachak-I, Ratua-I, and Manikchak.

Good road network makes the transportation system effective, and it is also found as a strength in Kaliachak-I, Kaliachak-II, and Chanchal-II block.

The high-value crops such as makhana, vegetables, oilseeds, fruits, betel vine, cocoon, and medicinal crops have been cultivating in some of the blocks like Harishchandrapur-I, Harishchandrapur-II, Kaliachak-I, Gazole, and Habibpur in the district. Localized and small mango-based industries are found in Kaliachak-III, Chanchal-I, and Harishchandrapur-II block. Because of the migration of the working labor force and opening up of small and medium shops in the district, the outsourcing of the farmers in Kaliachak-II and Manikchak has been increasing continuously. The plenty and cheap labor availability is an advantage in seven blocks of the district.

The presence of weekly markets in Kaliachak- III, Chanchal-I, Chanchal-II, Habibpur, and Old Malda is another strength. The district receives a reasonably good amount of monsoon rainfall and water which fills ponds and shallow wetlands. And it is also a strength in blocks like Kaliachak-II, Bamongola, Harishchandrapur-I, Harishchandrapur-II, Chanchal-I, and Chanchal-II in the district. The self-help group (SHG) is actively working as the receiver of credit and loan which is very common in Kaliachak-III, Manikchak, and Harishchandrapur-I blocks in the district. Each year, silt deposition (because of flood) is commonly identified in Kaliachak-II, Ratua-II, Harishchandrapur-I, the northern part of Gazole, and eastern part of Bamangola block. Small agro-industry such as rice mills are found in abundance in Kaliachak-III and Habibpur in the district. Use of organic manure and fertilizers has been considered as a strength in Kaliachak-III and English Bazar block.

The water retention capacity (because of clay soil) has support certain crops like paddy and fibers which is common in Kaliachak-I, Kaliachak-III, English Bazar, Harishchandrapur-I, Harishchandrapur-II, Gazole, Ratua-I, Ratua-II, and Manikchak

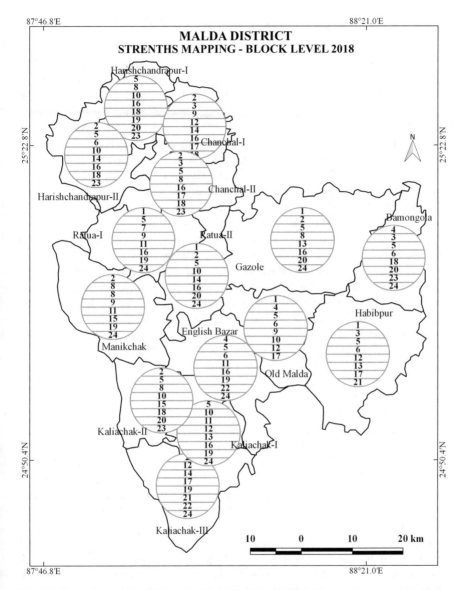

Fig. 6.1 Strengths mapping at block level in Malda District 2018. (Source: based on Table 6.3)

block in the district. The women participation in agriculture is high especially in vegetables and spice crops cultivation in Kaliachak-I, Kaliachak-III, Manikchak, English Bazar, Bamangola, Gazole, Ratua-I, and Ratua-II block in the district (Fig. 6.1).

6.2.2 Weaknesses Mapping in Malda District

The prominent weakness in agriculture of Malda is abundance of child labor. The term child labor (below the age of 14 years) is often defined as work that deprives children from their childhood, their potential, and dignity, and that is harmful to physical and mental development. In India, child labor is mainly attributed in agricultural sector (71%), 17% in services and 12% in industrial sector. As responded by ADO from Old Malda that the child labor is very common in cultivation process in Malda district. The detritus of soil quality has been experienced by the farmers as responded by the ADOs of Kaliachak-I, Kaliachak-III, Habibpur, Old Malda, and (Fig. 6.2). The heavy doses of pesticide are caused in decreasing quality of soil as well as crop loss especially in cereals and vegetables crops in Kaliachak-I, Kaliachak-III, Chanchal-II, and Harishchandrapur- II.

The high cost of cultivation is another important problem faced by all kinds (land size) of farmers. But, it is almost unbearable to the farmers having less than 0.20 hectares of land in the district. Because of the low socioeconomic status and small parcel of land size, the farmers have been facing problems in adaptation of new technological innovation for cultivating and processing the crops as it is responded to by the ADO of Bamangola. To boost up the economy from agriculture, there is lacking in number of agro-based industries as well as other industries which actually absorbs the working age-group population from agriculture as stated by the officials of Kaliachak-II, English Bazar, Gazole, Manikchak, Ratua-I, Ratua-II, and Harishchandrapur-II blocks in the district.

Access of goods and cold storage facilities for the farmers defiantly help to boost up income in the present and future which is essentially needed for potatoes, vegetables, and fruits. Till February 2018, there are only two public cold storages which failed to accommodate all types of crops. The response notes from four ADOs of the district. The method of cultivation is age-old because it is mostly practiced by the head of the households and therefore, the district is very slow in adaptation of the new farming system under different crops. The regulated market is essential to ensure the valid price of the agro products to the farmers. In the district, only two blocks have such a facility.

Population growth and family nucleation have resulted in fragmentation of land responded by five ADOs in the district. The yield and productivity of some crops like pulses, oilseeds, and fruits are low informed by the farmers of Bamongola, Manikchak, and Harishchandrapur-II. The lowering down groundwater table hampers negatively in irrigation sources reported from Bamangola, Habibpur, Old Malda, English Bazar, Gazole, and Kaliachak-I block in the district. It is seen in Chapter 3 that the district has more than 96 percent small and marginal farmers whose socioeconomic condition is not well enough to purchase and adopt new technologies and agricultural innovation replied by ADOs of Kaliachak-I, Gazole, Ratua-II, Harishchandrapur-I, Harishchandrapur-II, and Chanchal-I block in the district.

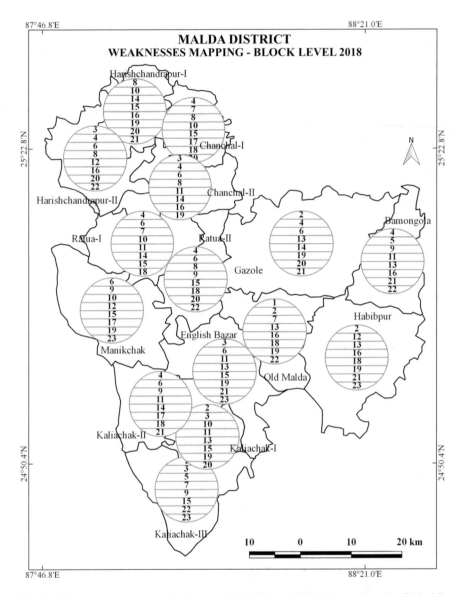

Fig. 6.2 Weaknesses mapping at block level in Malda District 2018. (Source: based on Table 6.3)

The heavy dosage of fertilizers has short-term gain, but in the long run, it reduces the natural quality of the soil. Lack in convenience of public transport (roadways and railways) facilities has also been identified as weakness to agriculture development as it is responded from Kaliachak-II, Bamangola, Habibpur, Old Malda, Manikchak, Chanchal-I, Chanchal-II, and Harishchandrapur-I block in the district.

More than 55 percent crop cultivating in the district is rain-fed which signifies that there is a limited irrigation source which is arising as a weakness in agriculture development reported from Habibpur, Old Malda, Chanchal-I, and Harishchandrapur-I blocks. Reducing government incentives is another weakness in agriculture answer records from six blocks. Slow adaptation of agriculture technology because of initial factor cost is another weakness in agriculture development as it is responded by five ADOs of the district. Paddy has a dominant share followed by wheat and jute which has been continued to grow by the farmers appearing in six blocks of the district.

The unscientific cultivation by the farmers is another important problem in low agricultural output which is observed in blocks like Kaliachak-III, Old Malda, Bamongola, Ratua-II, and Harishchandrapur-II. The remittances from productive age group migrants no doubt adds some amount in total family income, but it also left agriculture to the aged farmers who are mostly slow in the adaptation of agricultural innovations. Blocks like Kaliachak-III, English Bazar, Habibpur, and Manikchak have reported higher labor workforces' migration from the district.

6.2.3   Opportunity Mapping in Malda District

Because of different bio-ecological variations, the district has the potential of multifaceted opportunities to cater them as the growth engine of agriculture development.

Agriculture diversification is one of the dimensions to go with crops as well as livestock which promises better agricultural return. Four blocks, namely, Chanchal -I, II, Harishchandrapur-II, and Kaliachak-II have enough potential to go along with agriculture diversification in the district (Table 6.4). The agro-based industry has enough opportunity to absorb the workforce from different ages and social groups, as it is reported from eight blocks in the district. The crop-based industries from vegetables, fruits, and cereals have ample opportunities in the district. Crop diversification can be developed more at a micro-level to generate more income which is informed by English Bazar, Gazole, and Ratua-I block as number one opportunity, but in seven blocks, it is found as a significant opportunity.

Crop variety and their rotation is also an important opportunity to make the health of the soil better. Crop rotation also provides opportunity to make more income which is found in Manikchak, Old Malda, Ratua-I, Chanchal-II, and Harishchandrapur-I blocks (Table 6.4). Though the district receives a fair amount of monsoon rainfall but the water availability in the Barind region remains at significant problem almost nine months a year. For this reason, dry crop cultivation, especially in Bamangola, Gazole, Habibpur, Kaliachak-I, and Old Malda blocks, could be proven as a feasible and least cost farming option.

Cultivation of high-value crops draws higher income in Ratua-I, Bamangola, English Bazar, Chanchal-I, and Chanchal-II block in the district. It also promotes a demand-based product system for export. The existing agricultural system should be restructured because certain products from vegetables and mango (such as pickle) which already have the great place at national and international market must be maintained in its place which is reported from Kaliachak-I, English Bazar, Old Malda, and Ratua-I block.

As of now, the district has limited irrigation sources especially sprinkle and tank irrigation which save the water from unwanted wastage. Therefore, introduction of low water wastage techniques could be extended as an opportunity for an efficient irrigation system recommended by ADOs from Bamangola, English Bazar, Habibpur, and Old Malda in the district.

Export-oriented cultivation may have proven a possible option for agriculture development as per the replies from Harishchandrapur-I, Bamongola, Habibpur, Kaliachak-II, and Ratua-I block. To lift up livestock farming, fodder cultivation would be an important opportunity fixed by the officials of Kaliachak-III, Ratua-II, Manikchak, Gazole, Habibpur, and Chanchal-I block in the district (Fig. 6.3).

Though jute cultivation enjoys better seats but the number of jute industries are going down day by day. It is the high time to promote and extend the jute factory again informed by the officials of Chanchal-I block which is the largest producer of jute in the district. The opportunity for better agriculture is also rested to bring some fallow land under cultivation system especially in Gazole block. The extension of the regulated market assumes an important reason to ensure low price of agricultural products. Because of limited market infrastructure, farmers mostly sell their products from fields or homes which did not offer better price very often informed from 6 blocks. For cereals and pulse crops, the field processing unit is very low; therefore, it is generally taking a long time to process the crops informed by an ADO from English Bazar block.

There are plenty of opportunities to increase pulse, oilseed, spice, and fruit crop production which is still lower than the district's potential. To increase production and productivity of mentioned crops, the crop category-wise production can be targeted, responded from Ratua-I, Habibpur, Chanchal-, and Bamangola block. An ADO from Gazole block informed that agricultural extension services should be extended to promote productivity and better opportunity for the farmers. Though intensive cultivation is very much in practice, mix cultivation has the future over intensive cultivation, reported from 11 blocks in the district.

The opportunities are also rested on NGOs especially for the betterment of the poor farmers. It is reported from five blocks that to earn sustainable profit from agriculture, on time knowledge share and investment can be equipped only from NGOs. The urban access blocks like English Bazar, Kaliachak-I, and Old Malda have potential to develop an urban-based agricultural system at the block level by poor farmers. For that, public-private partnerships and involvement of NGOs are highly appreciated.

The agro-products of organic farming have present demand and promising future as it is replied by the ADOs of nine blocks in the district (Fig. 6.3). For that, plantation-based agriculture or mixed farming could be extended in the district, reported from Ratua-II, Manikchak, Kaliachak-III, and Gazole block. To make organic farming profitable, road and railway connectivity could be used in a more efficient way specially for agricultural marketing and export-import purpose answered by ADO of Kaliachak-II in the district.

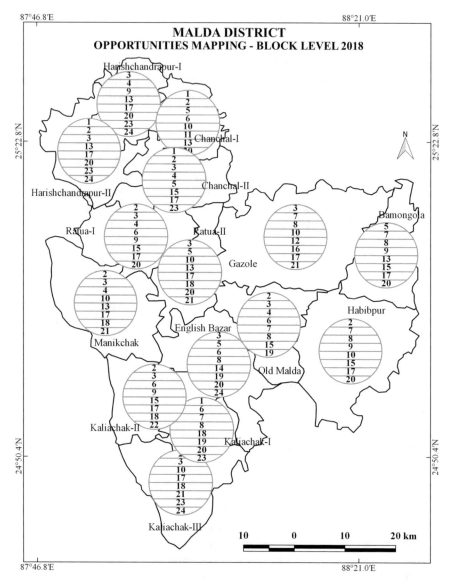

Fig. 6.3 Opportunities mapping at block level in Malda District 2018. (Source: based on Table 6.4)

The soil test-based nutrition management is necessary to reduce unwanted and excess use of fertilizers responded by the five ADOs (Fig. 6.2). To boost up district agriculture, multiple wetlands including ponds, small lakes, and swallows deep land can be targeted for pisciculture as well as wetland-based farming such as makhana and others informed by ADOs of English Bazar, Harishchandrapur-I, Harishchandrapur-II, and Kaliachak block of district.

6.2.4 Threats Mapping in Malda District

Agriculture in Malda district has been facing a number of problems. Some of them are very frequent and recurrent too.

Crop disease is very sporadic and unobvious across different crop categories responded by the officials of Chanchal-II, Habibpur, Harishchandrapur-I and Harishchandrapur-II in the district. Apart from unwanted crop disease, the decreasing trends in number of wetlands is a major concern before service extension planners for Harishchandrapur-II, Chanchal-I, Kaliachak-I, and Manikchak block. The water shortage in cultivation especially for vegetables and pulses is a major concern in ten blocks reported by the ADOs.

Erratic rainfall and its variability cause huge crop loss in the district as it is reported by the farmers as well as ADOs of nine blocks. The excessive use of fertilizers is causing soil fertility loss which is problematic in six blocks (Fig. 6.4). The flood causes loss of the standing crops, as well as agricultural land which is very apparent in five blocks along major river belts of the district. The fluctuation in supply of agricultural products and produce in the national market is another critical threat to the production system of the district reported from five blocks. The Bangladeshi immigrant is one of the threats at Bamangola block.

It is a fact that increasing temperature hinders plant growth which is being realized by the farmers as well as policy analysts of 11 blocks in the district. The excess dosage of fertilizer has caused many social and economic problems which are very much in practice in five blocks. Hailstorms and Kalbaishakhi are very common during the month of April and May every year. Hailstorms damage crops in the district (especially fruits, vegetables, and cereal crops).

Land encroachment for the sake of apartments and multistoried buildings is common in Old Malda, Habibpur, and Kaliachak-I block in the district. Due to this, each year agriculturally productive land is converted into concrete land. Bamangola block is affected by land security problems because of its bordered location. The district is going through inconsistency of rainfall. The blocks like Bamangola, Ratua- I, and Gazole are very much dependent on rainfall-based cultivation due to less accessibility of groundwater. On the other hand, because of the flood, excess water flow caused soil erosion in Kaliachak-III, Manikchak, and both Harishchandrapur every year.

Price instability and market volatility have caused a sudden fall of price which has been the major threat in eleven blocks of the district (Fig. 6.4). The farmers from six blocks have responded that the unexpected rainfall is a threat to them. During the winter season, mainly from the last 10 years, there is unexpected shadow fog that caused huge crop loss especially wheat and oilseeds in the district stated by two ADOs. Water scarcity remains a substantial problem in Old Malda and Bamangola blocks in district. On the other hand, water logging during monsoon season remains a major problem in the Tal region mainly in Harishchandrapur-I, Harishchandrapur-II and Chanchal-II blocks.

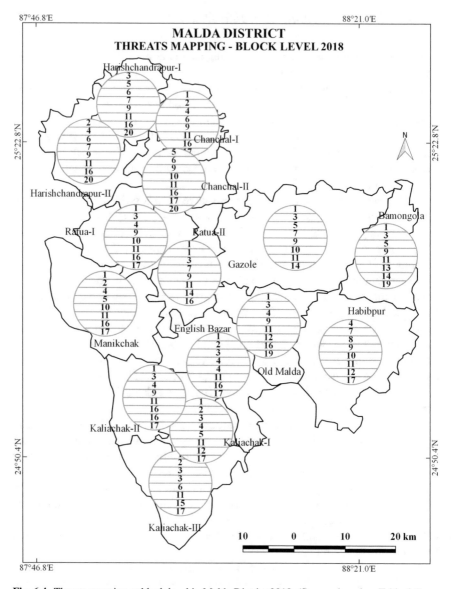

Fig. 6.4 Threats mapping at block level in Malda District 2018. (Source: based on Table 6.4)

6.3   Qualitative Matrix

Qualitative matrix is a method of analysis of qualitative information in to qualitative cum quantitative form in rows and columns. This method is extensively used in business and policy concerned research where information is obtained through either in-depth interview or focus group discussion (FGD). The major advantage of this method is that any qualitative information can be converted into quantitative form in either alphabetical or numerical or both simultaneously. This method also provides the base for priority-based policy intervention for local, regional, and national level policy. The major problems have been shortened through matrix method which is further processed in participatory research appraisal (PRA).

The basis of this method is rooted in 1988 (Gordon and Langmaid) as "cross-sectional, deductive, inductive and qualitative data analysis method in applied research." The matrix method (MM) has been altered from time to time, and it is formalized according to research approaches. Both in applied and academic research, MM has successfully answered many research questions on qualitative data (Groenland 2014). There are different steps and types of MM. For ease of research, creating a raw matrix, filling it, categorization, and tallying differsfrom single sample group to multi-sample groups in client-oriented research (Groenland 2016).

In the present study, in-depth interview is conducted with DDA (admin.) of Malda district in January 2018. The general objective of the interview was to know the agricultural situation in general and major problems faced by farmers of the district. The nature of problems has been identified mostly from local socioeconomic breakthroughs. Besides, the present agricultural situation of the district is influenced by many other external factors that arise in other districts, states, and even international markets. The existing and ongoing policies have been implemented in keeping view of national agricultural problems, and same have been generalized in the district. We have found in Chapter 1 that physical and human aspects of the study area at broader spectrum is somehow identical but not as same. The physical and human environment in the district have been greatly impacted on the agricultural situation.

In the first round of household survey and interaction with ADOs of different blocks, some of the major difficulties have been listed. Because multiple problems have been faced by the farmers, therefore firstly all problems have been listed (Table 6.6), and then top five problems of agriculture in the district are shortlisted through qualitative matrix.

After an extensive discussion, the top five problems like irrigation shortage, failure of minimum support price (MSP), lack of agro-industry, limited market opportunities, and lack of cold storage have been identified (Table 6.5). To prioritize the problem, the researcher has asked to select one out of two problems like between cold storage and irrigation; if farmers chose irrigation, then the same is noted in column number six of Table 6.5. And, the process is repeated with third, fourth, and fifth rows. Based on frequency of response, the column seven, i.e., "score," is calculated. And, finally, the rank has been assigned based on column seven.

From Table 6.5, it is found that failing of MSP arose as a number one problem in the district, followed by cold storage, irrigation facilities, markets, and agro-industry. At the block level, all top five problems have been arranged further in Table 6.6 in alphabetical order. The presentation of major problems has been displayed by land size class category wise at block level in the district.

Table 6.5 Qualitative matrix

A. Irrigation	B. MSP Failing	C. Agri-industry	D. Market	E. Cold Storage		Score	Rank
	B	A	A	E	A. Irrigation	2 A	3
		B	B	B	B. MSP Failing	4 B	1
			D	E	C. Agri industry	0 C	5
				E	D. Market	1 D	4
					E. Cold Storage	3 E	2

Source: processed by author, based on field survey 2018

Table 6.6 List of noted problems and their ranking in Malda District 2018

Rank	Descriptions	Rank	Descriptions
1	Cost of cultivation increasing	14	Land quality decreasing
2	Cost of production increased	15	Land size decreasing
3	Distant regulated market	16	Limited transportations
4	Electricity not ensured	17	Minimum support price (MSP) failing
5	Erratic rainfall	18	Productivity decreasing
6	Flood risk high	19	Productivity not good
7	Govt. incentive lacking	20	Profit decreasing
8	Input cost increased	21	Profit margin decrease
9	Irrigation shortage	22	Quality of seeds not ensured
10	Labor cost increased	23	Rainfall decreasing
11	Lack of cold storage	24	Rainfall variability
12	Lacking farm manager	25	Risk of production high
13	Lack of regulated market	26	Transport cost increased

Source: field survey 2018

6.4 Top Five Problems Under Different Land Size Classes

6.4.1 Top Five Problems Under Marginal Land Size Farmers in Malda District

In the marginal land size class category, the first problem of land size decreasing appeared in five blocks, namely, Chanchal-I, Chanchal-II, English Bazar, Harishchandrapur-I, and Kaliachak-I in the district (Fig. 6.5). The irrigation shortage and flood risk appear as the number one problem in Bamangola and Habibpur and Kaliachak-II and Ratua-II block, respectively. The cost of cultivation is noted as increasing in Gazole block. Decreasing profit margin in Harishchandrapur-II, limited transportation in Kaliachak-III, lacking farm manager in Manikchak, decreasing land quality in Old Malda, electricity not insured in Ratua-I have been noted as prime problems in the district.

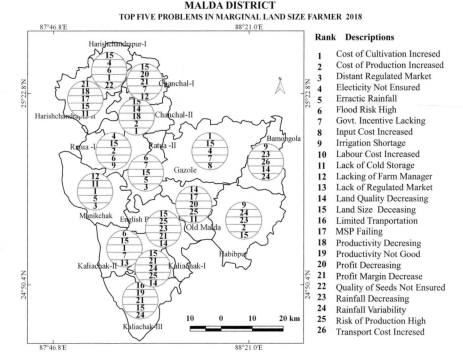

Fig. 6.5 Top five problems of marginal land size farmer in Malda District 2018. (Source: based on Table 6.7)

Decreasing land size is the second problem in Gazole and Kaliachak-II block while rainfall decreasing is reported in Bamangola block, profit decreasing in Chanchal-I and Kaliachak-I, and land quality decreasing in Chanchal-II block. The risk of production is high in English Bazar due to the nearness of the urban center because this municipal block allows foodstuffs from different parts of the district. The rainfall variability in Habibpur, electricity not ensured in Harishchandrapur-I, and productivity decrease in Harishchandrapur-II are some of the problems that have been identified as the second most problem. The second problem includes low productivity in Kaliachak-III, lack of cold storage in Manikchak, MSP failing in Old Malda, and reducing government incentive in Ratua-II block in the district.

The nature of problems ranked third in the district is rainfall decreasing in English Bazar and Habibpur block. The cost of cultivation is increasing in Kaliachak-II and Manikchak block. The transportation cost increased in Bamongola and profit margin decreased in Chanchal-II. The productivity decreases in Chanchal-II, electricity not ensured in Gazole, flood risk high in Harishchandrapur-I, and failing MSP in Harishchandrapur-II have also been identified as the third problem. The rainfall variability in Kaliachak-I, profit margin decreased in Kaliachak-III, profit decrease in Old Malda, cost of cultivation increased in Ratua-I, and land size decreasing in Ratua-I are some of the problems too categorized under the third problem (Fig. 6.5).

The number four problem under marginal land size group displays as the cost of production increases in Chanchal-II and Habibpur block. The government incentive is lacking in Gazole, Chanchal-I, and Kaliachak-II. The persistence rainfall variability is reported in Manikchak and Ratua-II. The decreasing land size in Harishchandrapur-II and Kaliachak-III and the risk of production are high in Kaliachak-I and Old Malda block under fourth rank problem. In Bamangola, decreasing land quality, profit margin decreased in English Bazar, cost of cultivation increase in Harishchandrapur-I, and flood risk in Ratua-I block are some of the fourth rank problems identified in the district.

The fifth problem under the marginal land size category is reported as a distant regulated market from Manikchak and Ratua-II block. The lack of cold storage facilities for Harishchandrapur-II and Old Malda block and decreasing land quality have been found in Kaliachak-I and English Bazar block. The rainfall variability is reported in Bamangola and Kaliachak-III blocks. Lack of a responsible farm manager in Chanchal-I and increasing cost of cultivation in Chanchal-II block have ascended as fifth problems. The input cost increase in Gazole, decreasing land size in Habibpur, quality of seed not insured in blocks like Harishchandrapur-I, lack of regulated market in Kaliachak-III, and irrigation shortage in Ratua-I have been reported as major problems in the district.

6.4.2   Top Five Problems Under Small Land Size Farmers in Malda District

The number one problem in the small land size group depicts the increasing cost of cultivation in Gazole and Ratua-II block (Fig. 6.6). The irrigation shortage in English Bazar and Bamangola, lack of cold storage in Chanchal-II and Kaliachak-III, and decreasing land size in Kaliachak-I and Chanchal-I are also identified as the first problems in the district. The profit margin decreased in Habibpur, MSP failing in Harishchandrapur-I, and labor cost increase in Harishchandrapur-I are prominent number one difficulties faced by mentioned land size farmers. The other key problems are noted as decreasing profit in Kaliachak, high flood risk in Manikchak, increased input cost in Old Malda, and high risk of production in Ratua-I block.

The second problem under small land size category arises out in Habibpur and Kaliachak-I as the quality of seeds is not ensured, erratic rainfall in Gazole and Ratua-II block, reducing government incentives in cultivation is reported from Kaliachak-III and Manikchak block. In Bamangola, there is rainfall variability in different cropping seasons. In Chanchal-I, decreasing land quality, the high risk of production (Chanchal-II) and increasing cost of cultivation (English Bazar) have been identified as foremost problems. From Harishchandrapur-I and Harishchandrapur-II, the problem is reported as declining profit and high flood risk, respectively. The distant regulated market (Kaliachak-II), lack of cold storage (Old

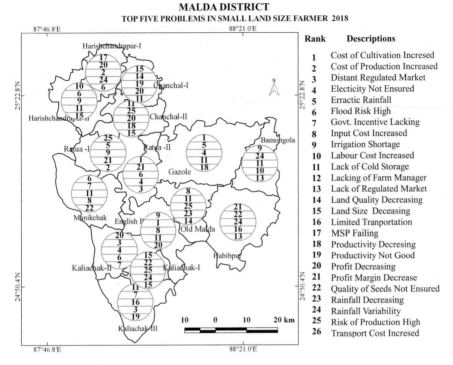

Fig. 6.6 Top five problems of small land size farmer in Malda District 2018. (Source: based on Table 6.7)

Malda), and decreasing profit margin (Ratua-II) have also been reported in the district.

The number three problem in different blocks appears as electricity which is not ensured in Gazole, Kaliachak-II, Harishchandrapur-II, and Ratua-I block. The transportation cost has increased in Kaliachak-I and Old Malda block. In Chanchal-I block, it is low productivity, while profit is decreasing reported in Chanchal-II block. The input cost increases in English Bazar, and rainfall variability is a visible problem in Habibpur block. The cost of production increases as compared to income, reported from Harishchandrapur-I. The limited transportation is a barrier for agriculture development in Kaliachak-III, and high flood risk is noted in Ratua-II block (Fig. 6.6).

The problem number four includes lack of cold storage in Harishchandrapur-II, English Bazar and Gazole block. Also, the rainfall variability causes crop loss in Harishchandrapur-I and Kaliachak-I block. Increasing labor cost in Bamongola, decreasing profit margin in Chanchal-I block as a result of decreasing productivity is number fourth problem reported in Chanchal-II block. In Habibpur block, it is limited transportation, high flood risk in Kaliachak-II, distant regulated market in Kaliachak-III, import cost increase in Manikchak, rainfall decreasing in Old Malda, while profit margin decrease in Ratua-I and electricity is not insured in Ratua-II block.

The fifth problem (land size decreasing) under small land size class category is common in Chanchal-II, Kaliachak, and Harishchandrapur-II blocks. Bamangola and Habibpur blocks report lack of regulated markets, Chanchal-I reports lack of cold storage, and profit decreasing and bad kind of productivity have been found as the fifth problem in English Bazar and Gazole blocks, respectively. The high flood risk, government incentive lacking, productivity not good, and quality of seed not insured are some of the difficulties found in Harishchandrapur-I, Kaliachak-I, Kaliachak-III, and Manikchak blocks in the district. In Old Malda, Ratua-I, and Ratua-II blocks, the decreasing of land quality, cost of production increase, and distant regulated market are some of the respective problems in the district.

6.4.3   Top Five Problems Under Semi-medium Land Size Class in Malda District

In semi-medium land size class, the number one problems are irrigation sources in four blocks, lack of cold storage in two blocks, and decreasing land size in two blocks in the district (Fig. 6.7). The decrease productivity, lack of regulated

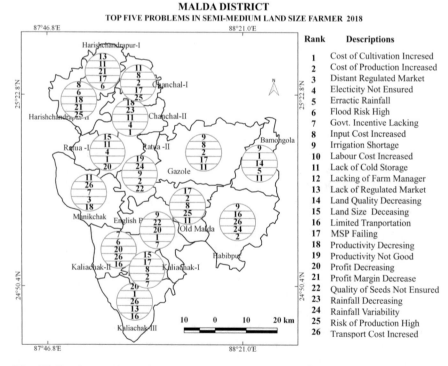

Fig. 6.7 Top five problems of semi-medium land size farmer in Malda District 2018. (Source: based on Table 6.7)

markets, and import cost increased are some of the problems identified in Chanchal-II, Harishchandrapur-I, and Harishchandrapur-II block correspondingly. In Kaliachak-II, III, Old Malda, and Ratua-II blocks, the problems like lacking government incentive, decreasing profit, failing MSP, and productivity not good are respective problems in the district.

The second problems which is obvious in two blocks are the cost of cultivation increasing, import cost increased, and lack of cold storage (Fig. 6.7). The decreasing rainfall, low quality of seed along with limited transportation, high flood risk, failing MSP, increasing transportation cost, rainfall variability, decreasing land quality, and high cost of production have been identified as the second problem under different blocks in the district.

The third problem includes decreasing profit, increasing cost of production, increasing inputs, and transportation which are revealed as typical problems in some of the blocks. The fourth problems such as increasing cost of cultivation, MSP failing, increasing cost of production are some of the problems noted in Fig. 6.7. The fifth problems such as lack of government incentives, lack of cold storage, and high risk of production have been found in different blocks of the district.

6.4.4   Top Five Problems Under Medium Land Size Farmers in Malda District

In the medium land class category, the number one problem arises at block level as high flood risk in two blocks, lack of cold storage in four blocks, and high risk of production that are reported in two blocks of the district. Some of the problems such as lack of regulated market, increase in the cost of production and transportation, lacking in government incentive, increasing input cost, and decreasing profit margin are also found as notable problems in some blocks (Fig. 6.8).

The second problem we found in different blocks as cold storage, lack of regulated markets, increase in the cost of production, failing MSP. Increasing cost of production appears very frequently in blocks like Harishchandrapur-I, Chanchal-I and Old Malda in the district.

The third common problems which looks more noticeable in the map are lack of government incentives, increase in transportation cost, failing MSP, and others. Increasing transport cost appeared in two blocks, i.e., Manikchak and Kaliachak-II.

The fourth problem includes decreasing profit, reducing government incentive, increasing cost of production, and decreasing profit margin which depicted maximum time under different blocks in the district.

The fifth or last common problem in different blocks is failing MSP and low productivity. In Chanchal-II and English Bazar, decreasing rainfall and increasing labor cost are the respective problems under different blocks. Harishchandrapur-II, Kaliachak-I, Kaliachak-II, Kaliachak-III, Manikchak, Old Malda, and Ratua-I and Ratua-II have been facing the problems of decreasing land quality, increasing cost of production, lacking government incentive, decreasing productivity, high risk of production, and increased transportation cost, respectively.

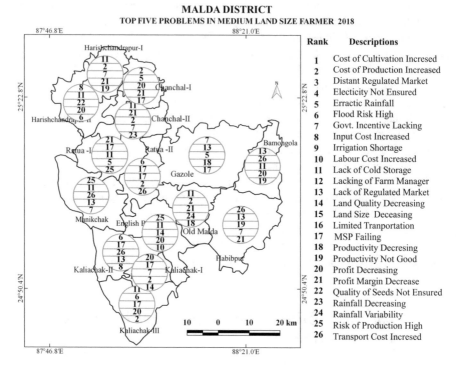

Fig. 6.8 Top five problems of medium land size farmer in Malda District 2018. (Source: based on Table 6.7)

6.4.5   Top Five Problems Under All Land Size Farmers in Malda District

The first problem under all land size classes includes decreasing land size in Chanchal-I, English Bazar, Gazole, Harishchandrapur-II, and Kaliachak-I block. The farmers report that failing MSP in Chanchal-II, Harishchandrapur-II, irrigation shortage in Bamangola, Habibpur, Kaliachak-III, and flood risk in Kaliachak-II and Manikchak block as major problems in agriculture practices. The decreasing profit, increasing cost of production, and lack of cold storage have been identified as vital problems in Old Malda, Ratua-I, and Ratua-II block in the district.

The second problem under all land size category is reported as increasing labor cost, increasing cost of production, irrigation shortage, increasing transportation cost, and decreasing land size under different blocks in the district. And other specific problems are reported as rainfall variability in Bamongola, lack of cold storage in Chanchal-I, decreasing land quality in Habibpur, and high risk of agricultural production in Kaliachak-II block in the district (Fig. 6.9).

The third critical problem under different blocks includes high flood risk and erratic rainfall during different cropping seasons in Habibpur, Kaliachak-III,

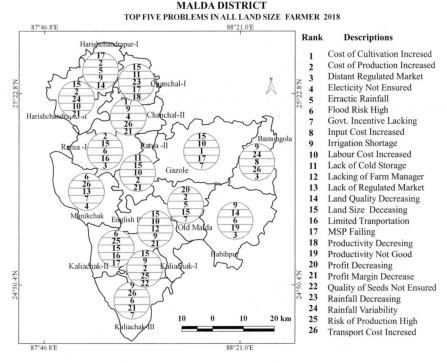

MALDA DISTRICT
TOP FIVE PROBLEMS IN ALL LAND SIZE FARMER 2018

Rank	Descriptions
1	Cost of Cultivation Incresed
2	Cost of Production Increased
3	Distant Regulated Market
4	Electicity Not Ensured
5	Erractic Rainfall
6	Flood Risk High
7	Govt. Incentive Lacking
8	Input Cost Increased
9	Irrigation Shortage
10	Labour Cost Increased
11	Lack of Cold Storage
12	Lacking of Farm Manager
13	Lack of Regulated Market
14	Land Quality Decreasing
15	Land Size Deceasing
16	Limited Tranportation
17	MSP Failing
18	Productivity Decresing
19	Productivity Not Good
20	Profit Decreasing
21	Profit Margin Decrease
22	Quality of Seeds Not Ensured
23	Rainfall Decreasing
24	Rainfall Variability
25	Risk of Production High
26	Transport Cost Incresed

Fig. 6.9 Top Five Problems of all land size classes farmer in Malda District 2018. (Source: based on Table 6.7)

Ratua-II, Harishchandrapur-I, and Old Malda blocks, respectively. The increasing cost of cultivation (Gazole) and production (Kaliachak-I) is also noted as the third problem in the district. The input cost increased (Bamongola), labor cost increased (Ratua-II), and lack of farm manager (English Bazar) have also been found as the third problem in the district. The lack of regulated market (Manikchak) and decreasing land size (Kaliachak-II) have also been reported as the third problem.

The fourth important problem is noted as irrigation shortage, limited transportations, failing MSP, and increased transportation cost in different blocks of district. Specific regional problems arise as low productivity in Habibpur, increased labor cost in Harishchandrapur-II, the decreasing profit margin in Kaliachak-III, lack of government incentive in Manikchak, decreasing land size in Old Malda, and increasing cost of production in Ratua-II block which are also considered as major hindrance to agricultural development in the district.

The fifth significant problem has been identified as a distant regulated market, decreasing profit margin, and lack of government incentive (Fig. 6.9). The decreasing productivity in Chanchal-I, decreasing land quality in Harishchandrapur-I, low quality of seed in Kaliachak-I, MSP failing in Kaliachak-II, electricity not ensured in Manikchak, and distant regulated market in Ratua-I block have been identified as fifth major problem in the district.

6.5   Participatory Rural Appraisal (PRA)

The participatory rural appraisal (PRA) is a method to know and describe rural problems, their knowledge conditions and to suggest, plan, and act on perspective solutions from viewpoints which is based on active participation of the local people. As a method, it includes baskets of techniques to acquire information about life and knowledge of rural people expands from rural resources, livelihood, social existence, community development, and many others. After the 1990s, the policy framework of India has changed from a "top-down" approach to a "bottom-up" approach. The nature of national policy has no more sketched from the top most governing body directly, especially those which relate to rural development; rather policy framing at primary level is decided by rural people according to their demand and requirement.

As a concept, PRA is emerged from a social science research approach, i.e., rapid rural appraisal (RRA) which was popularly used in 1970s. The RRA simply focuses on collecting, analyzing, and evaluating information on rural life and knowledge. In RRA, information has been generated from rural people in close cooperation with their adjusted conditions, literacy, and communication. The process and outcome of RRA from researcher's perspective is a comprehensive realization of needs and requirements of local folk. To overcome partial participation measures of rural folk, the PRA has developed. As Cavestro (2003) cited, Schoenhut and Kievelitz states "PRA is an extractive research methodology consisting of systematic, semi-structured activities conducted on-site by a multidisciplinary team with the aim of quickly and efficiently acquiring new information about rural life and rural resources." The attitude of the PRA process is to empower rural people after knowing their exclusive way of understanding and therefore "PRA is a growing combination of approaches and methods that enable rural people to share, enhance and analyze their knowledge of life and conditions, to plan and act and to monitor and evaluate. The role of the outsider is that of a catalyst, a facilitator of processes within a community which is prepared to alter their situation" cited by Cavestro (2003).

6.6   Application of PRA in Malda's Agriculture

There are many methods and approaches to participatory appraisal. In the present study, focus group discussion (FGD) has been used. After 1992, India's policy intervention process followed the bottom-up approach where the village community was given the opportunities to discuss their problem first. The summary of discussion with respect to problems and prospects of local people submits to the concerned authority of Community Development/Village Panchayat for further considerations. Before reaching the state or central government's planning commission, recommendations from community development officers were further deliberated in detail by respective departments at district headquarters. Here, necessary work on agriculture development is done by the Deputy Director of Agriculture (Admin.) after the consultation with the Agriculture Development Officer of different blocks. Present PRA and final processing have been done in the presence of ADOs and DDA (Admin) of the district in March, 2018 at DDA (Admin.) office, Malda.

Table 6.7 is an outcome of focus group discussion with DDA (admin.) and ADOs of the district. This table is based on both qualitative and quantitative information to sort out problems for specific policy execution for the agricultural development in the district. Since more than 96 percent farmers comprise small and marginal land size category, the above-discussed problem in the district is common and dominant under all land classes in the district. Here an attempt has been made to priorities policy intervention which fits for all land class size categories in the district. Since the problem identification is obtained from primary survey and their determination is followed on the qualitative matrix, therefore, the use of PRA technique is just appropriate in the present context.

In present PRA analysis, Robert Chamber's scale has been used in which 5 is the minimum number and 20 is the maximum. In the PRA (Table 6.7), five important aspects have been taken into consideration which have been followed by Indian policy execution committee. The nature of scale is discussed below;

1. In the table, the "long-term result" is a real positive scale which means to solve any selected problem; if it (suppose MSP) has long-term results, then it is likely to get higher marks, and if chosen particular has short-term result, then selected aspects will get lower marks.
2. The "less time to solve" is a negative scale, meaning thereby the higher the time it takes to explain, the lower the marks it would be and vice versa.
3. In case of involvement of the "amount of money to solve" a particular problem, if the fund requires high, then it will be scored low and vice versa and therefore again it bears a negative scale.
4. If the "scope of people participation" on a given particular is high, then the number would be high and vice versa. In this sense "scope for people participation" is a positive scale.
5. The "administrative hurdle" is another dimension to measure the policy implementation process. Here, if administrative hurdles are full paper works with long steps of procedure, then marks would be lower and vice versa, and that is why this is again a negative scale.

Based on Table 6.5, the given particular of Table 6.8 has been decided. Score and rank in the table have been arranged according to the maximum number secured by each particular in total hundred (100).

Table 6.8 shows that if the MSP falling problem is to be solved, then it requires a very less amount of money, but it would allow only high people participation. Since the MSP of crops is fixed by the state and central government, therefore, it also changes from time to time. It becomes a difficult task in terms of Administrative procedure. If the district suggests some specific MSP recovery policy, it requires a long time to implement, may be years; therefore, it secures only 10. The MSP

Table 6.7 Ranking of top five problems at block level in Malda District 2018

Block	Marginal					Small					Semi-medium					Medium					All classes				
Bamongola	9	23	26	14	24	9	24	11	10	13	9	1	14	5	11	13	26	11	20	19	9	24	8	26	3
Chanchal-I	15	20	21	7	12	15	14	19	20	11	11	8	2	17	25	2	5	20	21	17	15	11	23	17	18
Chanchal-II	15	14	18	2	1	11	25	20	18	15	18	23	11	4	1	11	21	2	7	23	17	9	4	26	21
English Bazar	15	25	23	21	14	9	1	8	11	20	9	22	20	1	7	25	11	14	20	10	10	10	12	9	21
Gazole	1	15	4	7	8	1	5	4	11	18	9	8	2	17	11	7	13	5	18	17	15	10	1	17	7
Habibpur	9	24	23	2	15	21	22	24	16	13	9	16	26	24	2	26	13	19	7	21	9	14	6	19	3
Harishchandrapur-I	15	4	6	1	22	17	20	2	24	6	13	11	21	17	6	11	2	7	21	19	17	2	5	9	14
Harishchandrapur-II	21	18	17	15	11	10	6	9	11	15	8	6	18	21	25	8	11	22	20	6	15	2	24	10	21
Kaliachak-I	15	21	24	25	14	15	22	25	24	15	15	17	8	2	7	20	17	7	2	14	15	9	2	25	22
Kaliachak-II	6	15	1	7	13	20	3	4	6	7	7	6	20	26	16	6	17	26	13	8	6	25	15	16	17
Kaliachak-III	16	19	21	15	24	11	7	16	3	19	20	1	26	13	16	11	6	17	20	2	9	26	6	21	7
Manikchak	12	11	1	5	3	6	7	11	8	22	11	26	7	3	18	25	11	26	13	7	6	26	13	7	4
Old Malda	14	17	20	25	11	8	11	25	23	14	17	2	8	25	11	11	2	21	24	18	20	2	5	15	7
Ratua-I	4	15	2	6	9	25	5	9	21	2	15	11	4	1	20	21	17	11	5	25	2	15	6	16	3
Ratua-II	6	7	15	5	3	1	21	6	4	3	19	24	9	2	22	6	17	17	2	26	11	15	10	2	21

Source: field survey, based on Table 6.6

Table 6.8 Participatory rural appraisal on top five problems in Malda District 2018. (**Weightage –** maximum: 20 minimum: 05 (Robert Chamber Scale))

Particular	Long term result	Less time to solve	Not much money to solve	Scope for people participation	Less administrative hurdles	Score	Rank
1. MSP falling	15	10	20	20	10	**75/100**	II
2. Cold storage	20	8	11	20	10	**69/100**	IV
3. Irrigation	20	12	9	15	18	**74/100**	III
4. Market	20	10	8	20	10	**68/100**	V
5. Agro-industry	20	12	14	20	13	**79/100**	I
TOTAL	**95/100**	**52/100**	**62/100**	**85/100**	**61/100**		

Source: calculated by the researcher based on field survey 2018 (see Table 6.5)
Note: marked by researcher in presence of DDA (Admin.) and ADO

revision and its solicitation especially in agriculture production no doubt has assumed a new boost, but it is only an important dimension and it's not the end. Therefore, its long-term effect scores 15. After looking into five criteria, the MSP scores 75 out of hundred (100).

The establishment of cold storage and spread facilities is very necessary especially for fruits and vegetables type crops. Till February 2018, the district has only two public run cold storages in which one is functional and one is not in working condition. No doubt access to cold storage will generate more income to the farmers; therefore, the long-term result of cold storage scores 20 out of 20. The problem of cold storage requires time and effort to resolve at high stakes; therefore, it secures only 8 out of 20. The setting up of cold storage and its maintenance required electricity as well as manpower which demands monetary investment. Therefore, it got 11 out of 20. The scope of people participation in vegetables and fruit cultivation no doubt will increase if the farmers get access to the benefit of cold storage. So, people participating in the production system report 20 out of 20. In the district, the establishment of cold storage comes under government concern only. Therefore, it is required to follow a number of administrative procedures, and that is why it secures 10 out of 20. The overall score on the cold storage facility secures 69 out of 100 in the district.

To provide ensured irrigation facilities to the farmers is an essential welfare duty of the government. To do so, many minor and major irrigation projects have to be introduced. No doubt ensured irrigation facilities have long-term results; therefore, it scores 20 out of 20. Different irrigation technologies and their setup is a time taking process, and that is why it scores 12 out of 20. The irrigation infrastructure of the district is traditional and not efficient in comparison with other agriculturally developed states' machineries such as Punjab and Haryana. To bring those in district agricultural system, again it requires money; therefore, it scores 9 out of 20.

The extension of irrigation will defiantly boost up more intensive agriculture. Therefore, it scores 15 out of 20. In terms of administrative hurdles, it scores 18, because the district has already sanctioned the plan for irrigation implements. The overall score for irrigation in the district is 74 out of hundred (100).

Till February 2018, the district has two regulated markets which assure MSP for different crops. Availability of markets is not uniform and easy to access because of distance and means of communication. To establish a market, it would have a long-term result in selling and buying the products at a reasonable price; therefore, it secures 20 out of 20. The state government has to pass the plan/bill in the assembly to set up a new regulated market for the district; it means it requires time to establish as well as monetary investment. Therefore, time scale secures 10 and amount scale gets 8 out of 20. Accessibility of the market will defiantly increase people's participation in transaction activity; therefore, it scores 20 out of 20. There is moderate to high administrative hurdle since it is an outcome of the planning process and that is why it secures 10 out of 20. The overall score of the market particular is 74 out of 100.

The district economy is based on agriculture and allied activity but there are very few agro-industries which are localized in nature. The Malda district is famous for mango, but the mango-based industry is not there. Still, a huge share of mango is being sold in the local markets therefore, mango growers earn less. The setting up of agro-industries, especially for fruits and vegetables will definitely have long term results, and therefore it scores 20 out of 20. The district has some planned expenditure to set up an agro-based industry, but because of an executional problem, it is not possible to set up. If the problem of agro-based industries is to be solved, then it will require not that much time. The establishment of an industry requires a huge amount of money, but the district has already planned this amount on this. In this case, the district has not to wait long for financial assistance from the government. The setting up of new industries will definitely generate employment; therefore, people participation would be very high. The administrative hurdle to establish industry is not that much difficult; therefore, it scores a moderate mark that is 13 out of 20. The overall score for the agro-industry is 79 out of hundred (100).

The overall analyses for priority-based policy intervention from Table 6.8 conclude that the agro-industry is scored maximum, i.e., 79 out of 100 and ranked first in the district. The MSP falling ranks second and scores 75 out of 100. The irrigation (74), cold storage (69), and market (68) have ranked three, four, and five respectively. After looking into five criteria, the specific priority-based policy intervention target would be agro-industry followed by MSP falling, irrigation, cold storage and market in the district. If policy to be targeted with respect to long-term results, in such a case cold storage, irrigation, market, and agro-industry would be the best choice to the planners.

With respect to time, the priority-based policy intervention would be agro-based industries and irrigation followed by MSP failing, market and cold storage. The money as a criterion if taken for priority-based policies intervention then MSP could be targeted at first and followed by agro-industry, cold storage, irrigation accordingly. The scope of people participation is if taken as policy intervention criteria then MSP, cold storage, market, and agro-industry can equally be targeted.

The administrative hurdle is less in case of irrigation followed by agro-industry and then market, cold storage, and MSP. The criteria-wise overall score for long term policy intervention shows that scope of people participation, money, administrative hurdle and time to solve have appeared in ascending order.

6.7 On-Going Programs/Schemes/Projects Running in Malda District

PRA analyses show that the setting up of agro-industries would be priority-based policy intervention for overall growth and development of the district. Table 6.9 shows that there are no such ongoing schemes, projects and programs related to agro-industrial setup or its development. Again, there is no such program directly related to cold storage, market and regular price revision for agricultural products in the district.

Although irrigation extension program is an indispensable part of Rastriya Krishi Vikas Yojana (RKVY), but the problem with this is, whatever the amount is received by the district, it is divided equally among the blocks to launch irrigation extension services. The study shows some blocks (Bamongola and Habibpur) in the district

Table 6.9 On-going programs/schemes/projects on agriculture development in Malda District 2018

Sl. No.	Name of on-going programs implemented in the district
1	Rastriya Krishi Bikash Yojana (RKVY)
2	Agricultural Technology Management Agency (ATMA)
3	National Food Security Mission (NFSM)
4	National Mission on Oilseeds and Oil Palm (NMOOP)
5	National Mission on Agil. Extension Technology (NMAet)
6	Rural Infrastructure Development Fund
7	National Mission for Sustainable Agriculture (NMSA)
8	Kishan Credit Card (KCC)
9	National Crop Insurance Programme (NCIP)
10	Bringing Green Revolution in Eastern India (BGREI)
11	A mission for Integrated Development of Horticulture (MIDH)
12	National Mission on Food Processing (NMFP)
13	Augmentation of meat production by intensive sheep, goat rearing (NMPS)
14	Optimization of productive efficiency through parasitic control
15	Distribution of inputs for improvement of livelihood through goat farming
16	Bishesh Go Sampad Bikash Avijan

Source: CDAP XII Plan, Malda, 2015

have been facing acute shortage of drinking water and irrigation which immediately requires priority-based intervention. So, here is the policy suggestion to RKVJ that the amount of money should be dispersed according to the severity of the problem rather than equality as a criterion.

References

Cavestro, L. (2003). *PRA-participatory rural appraisal concepts methodologies and techniques.* Padova: Padova University.

Groenland, E. (2014). The problem analysis for empirical studies. *International Journal of Business and Globalisation, 12*(3), 249–263.

Groenland, E. (2016). Using the matrix method for the analysis of deductive, qualitative research data. An introduction with an annotated illustration. *Qualitative Research Data.* (April 25, 2016).

Summary, Suggestions, and Conclusion

Summary

The summary from analytical chapters of the study has been depicted in following sequences:

Chapter 2 is on land use and cropping pattern dynamics. It is found that the net sown area, net cropped area, and total area of the district have decreased from 1995–1996 to 2015–2016. Although, net sown area has declined but gross cropped area increased across different land size categories in the district. Gross cropped area under cereals, pulses, and vegetables has noted negative growth, while fiber, oilseed, and fruit area increased across land size classes, but the maximum gain is noted under marginal land size class. The total area under food crops is decreased substantially over non-food crops. The amount of change in cropping pattern in land size classes is found that marginal category records the maximum followed by semi-medium and small group. Primary data reveals that profit, consumption, climate change, and productivity are the major reasons in cropping pattern change across land size classes in the district.

Agricultural development with respect of inputs and outputs has been explained in Chap. 3. It is noted that except temperature index, other physical input indexes such as soil nutrients, rainfall, and humidity have been decreasing over the years. The composites index of non-physical index shows the increasing trends in irrigation, seeds, mechanizations, credits, and socioeconomic aspects in the districts. The weightage index reveals that agriculture outputs (Crop diversification and productivity) explain about 32 percent of physical inputs while non-physical inputs describe 45 percent out of 50 percent in each input. The non-physical inputs are significant and positively correlated with crop diversification in which marginal category posits robust association. The overall agriculture development under different land size categories is found that overall improvement in index value in 2015–2016 over 1995–1996 with better pace in marginal category due to intensification of land use contributed by irrigation, HYV seeds, mechanization, fertilizer consumption, and improving socioeconomic conditions.

The positive and significant relation between crop diversification and agricultural development has proved that crop diversification increases along with the pace of agriculture development. Crop diversification and its status is explained separately in Chap. 4. Trends of crop diversification by land size classes at block level revealed the booming picture, and again it is leading in marginal category. In terms of crop category-wise diversification change, the pulses, spices, oilseeds and fibers are going down but cereals, vegetables, and fruits crop diversification increases in the district. The total number of crops under vegetables rises the maximum followed by spices, oilseeds, fruits and cereals under different land size classes. An insight from field survey illustrates that profit, consumption, productivity, and climate change are some of the reasons backed in decision-making of crop diversification in the district.

In Chap. 5, the agriculture development and crop diversification along with factors and relative efficiency of development have been examined in sample villages. It is found that agriculture development and crop diversification has established casual relation across land size classes. The eight indicators under three components has explained 78.5 percent of total variance in marginal land size class, 68.6 percent in small, 61.2 percent in semi-medium, 71.9 percent in medium, and 72.8 percent in all land size classes in the district. Agriculture development in sampled villages differs within and between land size categories. With respect of input-output ratio, Rangaipur, Jot Basanta, and Hazaratpur are efficient in marginal category meaning that thereby in same amount of inputs, these villages received better outputs than rest of the villages in the district. There are two villages in small land size category, three in semi-medium, two in medium and four in all land size classes which have been recognized as efficient villages in the district. It is also proved that the agricultural development not ensured efficient agriculture.

After going through wide-ranging analyses and explanation, an attempt has been made to look into the problems and prospect of agriculture development in Chap. 6. Although the district is passing through the problems like ensured irrigation, lack of agro-industries, MSP failing, market and cold storage, flood and hailstorms, price instability, decreasing soil nutrients, and others, it also has the potential and opportunities of crop diversification, agriculture diversification, organic farming, development of agro-industries, and wetland-based cultivation. The solution of existing problems has attempted to solve through PRA which shows that the agro-industry is most required option to rural folk. To gain the maximum outputs from crop diversification, setup of agro-industry would be the priority-based policy intervention in the district. Followed on same line, MSP failing, irrigation, cold storage, and market are subsequent policy particulars are needed immediate attentions. Since here PRA works with and for rural folks, therefore, its outcome relates to welfare and development for the farmers by the policy makers, administrators, and academician to fulfill the dream of welfare state.

Suggestions

Chapter 6 is the core of the study in terms of suggestive measures which should be executed for crop diversification and agricultural development. Apart from that, some other efforts could be consider which are noted in the following sequence:

1. It is good to change cropping pattern regularly; since it is dominated in marginal land size class, therefore, other land size categories should be encouraged into change from conventional to non-conventional cropping system. Here role of district agencies and district extension service center of agriculture and agriculture development officers have essentially been felt.
2. After realizing the threat from increasing temperature, soil humidity and climate vulnerability, the area extension program under dry farming and low water resistance crops should be extended through policy linkage with government of India.
3. Non-physical inputs, especially credit facilities, tank and sprinkle irrigation, HYV seeds, cold storage and market, and crop insurance, are also suggested to district administrator to implement and expand these with government of India's schemes.
4. Agriculture development and its forward linkage with crop diversification has also been felt in the study. Therefore, for the focused crop diversification, both physical and non-physical aspects of development should be taken care with respect to ongoing programs.
5. For efficient agriculture development, efficient villages could be taken as model village to other inefficient villages. To do it at macro level, it needs further research and knowledge sharing through sponsored projects and schemes from respective department of the district.

Conclusion

Essence of the present study on crop diversification and agriculture development has been felt because of two reasons: first, crop diversification ensures better output with promising future, and second, sustainability of agriculture development can be enhanced through crop diversification since it is natural process with minimal human intervention.

What we are seeing today in the agriculture of Malda is a product of many years of interactive process between man and nature. Today, there are many opportunities to revamp the sector and make it more productive. Internal opportunities include establishments of agro based industries, agriculture diversification, crop diversification, multiple cropping, organic farming, and linkage of outsourcing from other sectors to agriculture. External opportunities include the political and economic organizational frameworks from the Indian Council of Agricultural Research (ICAR) and its off-shoots, New Partnership for District Development and Comprehensive Malda's' Agricultural Development Program. In addition to it,

Malda is one of the National Planned Development districts, and it stands to provide the economic packages which can be used for better agricultural infrastructure and marketing. By doing this, agricultural outputs can gain national and international fame again.

The government has a very important role in promoting efficient and effective environment for agriculture development and crop diversification. For that, holistic approaches are needed in the identification, testing, and promotion of input and output aspects of agriculture. There are no one-size-fits-all solution. Neither the government nor the private sector can do it alone.

Ancient philosopher said "a man chases two rabbits catches neither." In the study, it is found that farmers are actually growing too many crops under different crop categories depending upon their suitability or profit opportunities. Here farmers are actually chasing too many rabbits. Heavy doses of fertilizers, insecticides, and pesticides, climate change, decreasing soil quality, price instability, volatility of market price, slow down public investment in agriculture, marginalization of faming systems, and poor socioeconomic conditions have also been attempted by the government to engaged and work with. Effort on the major policies related to mentioned issues, the governments have also been chasing too many rabbits which are too far, too long, to achieved. Now, this is the time to stop, to prioritize the focused development of agriculture in order to ensure holistic development of the district. It's a tradition in Asia called a word of wisdom; I am totally optimistic and do the same here.

Annexures

Annexures: Chapter 2

Annexure 2a: Crop listing

1995–1996		2015–2016			
Cereals	**Total fiber**	**Cereals**	Coriander	Litchi	Other tuber crops
Wheat	Jute	Bajra	Fennel	Mandarin orange	Other vegetables
Paddy	Cotton	Barley	Fenugreek	Mango	Peas
Maize	Mesta	Jowar	Green chillies	Miscellaneous fruits	Potato
Barley	Other fiber	Maize	Garlic	Other citrus	Pumpkin
Small millet	**Total oilseed**	Other cereal	Ginger	Papaya	Radish
Other cereals	Coconut	Paddy	Large cardamom	Pineapple	Spinach
Total fruits	Sesamum til	Ragi	Other condiments	Sapota	Sweet potato
Banana	Rapeseeds & mustard	Wheat	Radhuni	Guava	Tomato
Mango	Other oilseeds	**Pulses**	Red chili	Temperate fruits	Yam

(continued)

1995–1996		2015–2016			
Orange	Niger seeds	Gram	Red yellow chili	**Vegetables**	
Other fruits	Linseeds	Horse gram	Turmeric	Beans	
Total spices	**Total pulses**	Masur	**Oilseeds**	Bitter gourd	
Chilies	Tur	Moong	Castor seed	Bottle gourd	
Ginger	Masur	Other pulse	Coconut	Brengle	
Cardamom	Gram	Tur	Groundnut	Cabbage	
Other condiments	Other pulses	Urad	Linseed	Capsicum	
Total vegetables		**Fibre**	Rapeseed & mustard	Carrot	
Onion		Cotton	Safflower	Cauliflower	
Potato		Jute	Sesamum til	Colocasia	
Other vegetables		Mesta	Soybean	Cucumber	
		Sun hemp	Sunflower	Drumstick	
		Spices	**Fruits**	Elephant foot yam	
		Beetle nut	Banana	Lady finger	
		Black cumin	Guava	Onion	
		Black pepper	Jackfruit	Other gourds	

Source: Agriculture Census (from Malda only)

Annexure 2b: Percentage of gross cropped area in Malda District, 1995–1996 and 2015–2016

	Bam.	Ch-I	Ch-II	E.B.	Gaz.	Hbb.	HCP-I	HCP-II	KL-I	KL-II	KL-III	Manik.	Odl M	Rat.-I	Rat.II	District
1995–1996																
Marginal	32.83	44.72	50.58	40.82	32.39	32.82	55.69	39.52	58.83	51.64	35.80	57.32	40.97	33.71	37.06	41.32
Small	27.71	33.28	27.43	28.19	33.17	34.82	20.05	39.53	15.74	19.13	23.67	21.21	32.31	18.81	31.93	28.53
Semi-medium	26.50	17.32	16.78	17.83	24.42	22.37	16.12	17.09	17.01	15.26	25.31	12.88	16.46	28.12	23.44	20.30
Medium	12.58	4.69	5.21	12.50	9.76	9.52	7.96	3.59	8.01	13.96	14.83	8.60	10.02	19.10	7.57	9.61
Large	0.38	N.A	N.A	0.65	0.26	0.47	0.18	0.27	0.41	N.A	0.39	N.A	0.24	0.26	N.A	0.24
All class	100	100	100	100	100	100	100	100	100	100	100	100	100	100	100	100
2015–2016																
Marginal	53.71	62.89	56.37	54.00	55.66	45.99	64.72	59.07	68.17	62.98	63.06	59.79	57.64	60.52	60.63	58.00
Small	27.96	23.88	29.33	28.58	27.30	29.46	23.56	25.65	24.90	20.98	21.62	33.73	28.86	26.31	25.79	26.88
Semi-medium	13.86	11.81	13.18	14.69	15.05	17.55	11.14	14.83	6.34	15.40	13.33	5.56	11.26	11.39	12.64	13.00
Medium	4.30	1.42	1.12	2.69	1.98	6.98	0.58	0.43	0.57	0.64	1.89	0.72	2.22	1.77	0.93	2.09
Large	0.18	N.A	N.A	N.A	N.A	N.A	N.A	N.A	N.A	N.A	0.10	N.A	N.A	N.A	N.A	0.04
All class	100	100	100	100	100	100	100	100	100	100	100	100	100	100	100	100

Annexure 2c: Percentage of GCA under cereals in Malda District, 1995–1996 and 2015–2016

	Bam.	Ch-I	Ch-II	E.B.	Gaz.	Hbb.	HCP-I	HCP-II	KL-I	KL-II	KL-III	Manik.	Odl M	Rat.-I	Rat.II	District
1995–1996																
Marginal	73.72	69.54	74.32	65.94	73.95	86.18	81.46	79.03	34.70	35.68	56.70	60.82	82.31	63.48	87.22	70.26
Small	75.54	68.45	74.14	63.81	77.46	85.78	80.83	78.56	48.03	50.48	60.33	61.10	86.00	60.98	87.37	74.25
Semi-medium	80.69	71.99	72.25	63.46	78.45	81.76	88.29	79.32	49.89	60.77	58.41	62.80	86.18	57.85	90.43	73.58
Medium	82.83	74.98	78.89	66.55	79.26	81.63	77.84	67.79	69.31	57.45	57.20	69.43	85.88	62.18	79.29	72.21
Large	94.34	N.A	N.A	92.04	89.94	95.43	100.00	100.00	96.30	N.A	51.61	N.A	100.00	100.00	N.A	91.63
All class	77.30	69.86	74.16	65.14	76.77	84.66	82.18	78.55	42.41	45.38	58.05	61.87	84.54	61.27	87.42	72.31

(continued)

2015–2016

	Bam.	Ch-I	Ch-II	E.B.	Gaz.	Hbb.	HCP-I	HCP-II	KL-I	KL-II	KL-III	Manik.	Odl M	Rat.-I	Rat.II	District
Marginal	67.64	58.22	55.39	40.49	75.74	74.63	72.54	73.72	16.36	43.37	47.76	38.41	87.80	56.19	71.58	60.69
Small	69.81	60.19	57.29	37.99	82.14	74.48	77.22	71.19	33.95	70.62	51.00	44.79	92.45	61.31	69.07	64.72
Semi-medium	70.84	66.29	63.50	47.01	83.69	77.01	80.46	80.89	44.98	41.64	56.35	40.71	91.07	59.93	60.90	68.38
Medium	72.82	43.70	65.10	59.01	80.32	83.45	75.58	65.16	55.42	60.21	50.42	45.65	65.75	65.62	81.87	72.24
Large	84.24	N.A	N.A	N.A	N.A	N.A	N.A	N.A	N.A	N.A	32.17	N.A	N.A	N.A	N.A	60.44
All class	68.95	59.44	57.12	41.25	78.78	75.62	74.54	74.10	22.78	48.93	49.64	40.70	89.03	58.13	69.68	63.01

Annexure 2d: Percentage of GCA under pulses in Malda District, 1995–1996 and 2015–2016

1995–1996

	Bam.	Ch-I	Ch-II	E.B.	Gaz.	Hbb.	HCP-I	HCP-II	KL-I	KL-II	KL-III	Manik.	Odl M	Rat.-I	Rat.II	District
Marginal	0.07	2.04	0.55	10.90	1.66	0.43	0.27	1.01	1.40	6.65	13.84	17.46	1.20	15.11	6.56	5.20
Small	0.00	2.57	0.71	12.21	1.21	0.66	0.36	1.89	3.17	7.28	13.42	16.44	0.29	18.45	6.45	4.52
Semi-medium	0.00	1.54	0.68	10.01	1.50	1.20	0.30	1.10	8.77	1.26	16.13	15.72	0.35	19.73	2.15	5.39
Medium	0.00	1.04	0.00	9.03	1.58	4.78	0.00	3.57	2.74	16.18	11.92	13.58	0.94	13.03	6.87	6.47
Large	0.00	N.A	N.A	0.00	0.00	0.00	0.00	0.00	0.00	N.A	25.81	N.A	0.00	0.00	N.A	2.11
All class	0.02	2.08	0.59	10.81	1.46	1.09	0.27	1.46	3.03	7.28	14.08	16.69	0.74	16.60	5.51	5.16

2015–2016

	Bam.	Ch-I	Ch-II	E.B.	Gaz.	Hbb.	HCP-I	HCP-II	KL-I	KL-II	KL-III	Manik.	Odl M	Rat.-I	Rat.II	District
Marginal	0.43	13.97	5.34	2.41	0.05	0.01	0.44	12.68	0.40	1.05	1.59	6.24	0.07	7.35	1.96	4.15
Small	1.01	13.36	6.15	1.04	0.08	0.03	0.59	16.60	0.07	0.25	2.67	4.19	0.03	5.62	0.71	3.90
Semi-medium	0.11	11.22	5.89	0.63	0.13	0.06	0.02	6.04	0.83	0.05	2.69	2.44	0.04	2.19	0.14	2.31
Medium	0.43	0.14	0.00	2.16	0.52	0.28	1.91	1.45	3.38	13.43	5.24	9.40	1.90	7.69	1.99	1.63
Large	0.00	N.A	N.A	N.A	N.A	N.A	N.A	N.A	N.A	N.A	26.48	N.A	N.A	N.A	N.A	2.87
All class	0.55	13.30	5.59	1.75	0.08	0.04	0.44	12.66	0.36	0.81	2.06	5.35	0.10	6.31	1.40	3.79

Annexure 2e: Percentage of GCA under spices in Malda District, 1995–1996 and 2015–2016

	Bam.	Ch-I	Ch-II	E.B.	Gaz.	Hbb.	HCP-I	HCP-II	KL-I	KL-II	KL-III	Manik.	Odl M	Rat.-I	Rat.II	District
1995–1996																
Marginal	0.01	0.01	0.01	0.10	0.19	0.01	0.02	0.00	0.00	0.00	0.24	0.68	0.12	0.00	0.07	0.12
Small	0.00	0.56	0.00	0.82	0.05	0.02	0.00	0.00	0.00	0.00	0.20	0.42	0.07	0.00	0.00	0.14
Semi-medium	0.00	0.00	0.00	0.00	0.00	0.77	0.19	0.00	0.04	0.00	0.00	0.31	0.11	0.13	0.00	0.12
Medium	0.00	0.13	0.00	0.00	0.09	0.07	0.00	0.00	0.00	2.31	0.00	0.00	0.04	0.00	0.00	0.13
Large	0.00	N. A	N. A	0.00	0.00	0.00	0.00	0.00	0.00	N. A	0.00	N. A	0.00	0.00	N. A	0.00
All class	0.00	0.19	0.00	0.23	0.02	0.01	0.00	0.00	0.00	0.00	0.05	0.09	0.02	0.00	0.00	0.04
2015–2016																
Marginal	0.19	0.29	0.06	1.74	0.01	0.00	0.13	0.33	1.16	0.02	0.72	0.77	0.08	0.11	0.09	0.32
Small	0.00	0.03	0.00	1.05	0.00	0.00	0.05	0.03	0.00	0.00	0.59	0.01	0.01	0.00	0.00	0.09
Semi-medium	0.00	0.00	0.00	0.00	0.03	0.08	0.01	0.00	0.00	0.00	0.19	0.75	0.01	0.01	0.00	0.03
Medium	0.00	0.02	0.00	0.00	0.03	0.01	0.00	0.00	0.00	1.92	0.00	0.00	0.02	0.00	0.00	0.03
Large	0.00	N. A	N. A	N. A	N. A	N. A	N. A	N. A	N. A	N. A	0.00	N. A	N. A	N. A	N. A	0.00
All class	0.10	0.19	0.03	1.24	0.00	0.00	0.10	0.20	0.79	0.01	0.60	0.49	0.04	0.07	0.05	0.21

Annexure 2f: Percentage of GCA under oilseeds in Malda District, 1995–1996 and 2015–2016

	Bam.	Ch-I	Ch-II	E.B.	Gaz.	Hbb.	HCP-I	HCP-II	KL-I	KL-II	KL-III	Manik.	Odl M	Rat.-I	Rat.II	District
1995–1996																
Marginal	12.08	1.53	3.65	0.68	9.34	3.69	4.18	0.82	0.50	4.39	1.35	0.48	1.33	0.97	0.06	3.17
Small	12.73	2.44	4.30	0.39	8.03	4.11	6.52	0.37	1.30	3.84	1.64	0.73	1.51	1.22	0.00	3.67
Semi-medium	8.63	2.51	5.38	1.42	7.55	6.37	1.57	0.86	2.09	5.67	1.32	0.51	0.60	0.75	0.00	3.64
Medium	9.27	3.06	1.65	0.00	8.64	3.83	8.30	1.26	0.00	0.00	0.51	0.43	1.52	0.59	0.62	3.15
Large	0.94	N. A	N. A	41.79	34.91	40.61	25.00	0.00	1.85	N. A	33.33	N. A	39.62	14.74	N. A	53.04
All class	10.94	2.07	4.01	0.64	8.38	4.43	4.55	0.66	0.86	3.87	1.28	0.53	1.29	0.88	0.07	3.40

(continued)

2015–2016

	Bam.	Ch-I	Ch-II	E.B.	Gaz.	Hbb.	HCP-I	HCP-II	KL-I	KL-II	KL-III	Manik.	Odl M	Rat.-I	Rat.II	District
Marginal	25.15	5.86	9.11	9.09	11.69	19.55	8.08	4.72	3.91	0.23	4.52	7.89	4.19	7.27	7.29	9.29
Small	24.28	7.08	9.11	4.46	11.09	21.17	5.57	1.54	3.24	0.13	1.78	7.71	1.41	4.80	5.50	8.98
Semi-medium	26.10	4.49	8.20	10.12	10.90	17.51	5.34	1.09	2.68	0.21	2.48	14.20	2.12	4.61	4.80	9.05
Medium	22.55	6.61	1.68	2.18	5.86	13.00	7.48	0.51	7.12	4.48	4.34	0.27	3.28	4.18	0.18	9.62
Large	15.37	N.A	N.A	N.A	N.A	N.A	N.A	N.A	N.A	N.A	12.55	N.A	N.A	N.A	N.A	21.07
All class	24.91	6.00	8.91	7.73	11.29	19.21	7.18	3.35	3.68	0.23	3.65	8.11	3.14	6.26	6.44	9.18

Annexure 2g: Percentage of GCA under fibers in Malda District, 1995–1996 and 2015–2016

	Bam.	Ch-I	Ch-II	E.B.	Gaz.	Hbb.	HCP-I	HCP-II	KL-I	KL-II	KL-III	Manik.	Odl M	Rat.-I	Rat.II	District
1995–1996																
Marginal	0.05	17.90	12.11	0.69	3.44	0.24	11.30	9.76	0.39	0.32	16.47	7.64	1.44	8.97	2.05	6.88
Small	0.00	18.14	11.54	0.00	3.19	0.06	10.47	10.12	2.36	3.55	13.19	7.94	0.86	8.58	1.84	6.18
Semi-medium	0.00	15.64	11.26	0.35	2.06	0.19	8.95	9.43	1.78	0.00	13.46	7.41	1.44	11.29	1.56	5.72
Medium	0.00	9.12	5.54	0.00	2.00	0.15	13.86	15.67	0.00	0.10	13.96	4.31	1.25	12.83	4.05	5.61
Large	0.00	N.A	N.A	0.00	2.37	0.00	0.00	0.00	0.00	N.A	0.00	N.A	0.00	0.00	N.A	0.35
All class	0.02	17.17	11.47	0.34	2.88	0.16	10.94	10.03	0.90	0.86	14.57	7.39	1.23	10.26	2.02	6.31
2015–2016																
Marginal	2.43	18.95	2.74	8.78	3.12	1.64	16.79	7.10	8.50	20.24	39.79	31.92	1.42	19.73	8.26	11.99
Small	1.84	15.26	1.23	11.27	1.71	1.75	14.82	8.27	14.77	13.34	41.59	26.85	0.88	18.03	10.16	10.27
Semi-medium	0.78	15.78	1.02	4.01	0.78	2.41	11.84	8.01	0.17	10.08	35.90	23.06	0.56	21.77	9.92	8.09
Medium	2.44	2.33	12.71	4.22	1.37	0.80	14.55	28.13	4.07	0.08	29.20	3.61	6.23	15.98	13.31	5.36
Large	0.00	N.A	N.A	N.A	N.A	N.A	N.A	N.A	N.A	N.A	15.59	N.A	N.A	N.A	N.A	7.42
All class	2.03	17.46	2.19	8.66	2.35	1.75	15.76	7.62	9.50	17.10	39.44	29.49	1.27	19.45	9.00	10.88

Annexure 2h: Percentage of GCA under vegetables in Malda District, 1995–1996 and 2015–2016

	Bam.	Ch-I	Ch-II	E.B.	Gaz.	Hbb.	HCP-I	HCP-II	KL-I	KL-II	KL-III	Manik.	OLD M	Rat.-I	Rat.II	District
1995–1996																
Marginal	13.96	8.94	9.13	11.99	11.03	9.46	2.75	9.30	6.55	7.43	10.05	8.67	13.54	10.37	1.32	9.01
Small	11.73	7.71	9.10	8.76	9.35	9.36	1.81	8.97	8.32	6.28	8.73	8.70	11.27	9.48	0.57	8.42
Semi-medium	10.68	8.22	8.96	8.76	9.50	9.71	0.71	8.88	6.28	6.25	5.83	7.93	11.29	8.06	1.75	8.01
Medium	7.90	9.06	10.13	11.58	8.36	9.53	0.00	11.71	11.71	12.94	9.78	8.49	10.37	8.31	1.07	8.69
Large	0.00	N.A	N.A	0.00	2.37	0.00	0.00	0.00	0.00	N.A	18.28	N.A	0.00	0.00	N.A	1.85
All class	11.68	8.41	9.14	10.42	9.82	9.47	2.01	9.16	7.18	7.80	8.61	8.56	12.09	9.13	1.16	8.60
2015–2016																
Marginal	3.32	2.18	3.75	10.27	4.04	4.16	1.77	1.39	6.07	5.19	5.32	6.19	5.54	3.95	2.20	3.97
Small	2.64	1.87	2.81	7.46	1.95	2.58	1.68	2.29	0.18	2.99	1.86	1.32	5.19	2.33	0.65	2.44
Semi-medium	2.04	1.61	3.26	13.77	3.14	2.98	1.62	1.30	0.59	0.23	1.60	0.96	5.59	1.20	2.31	2.92
Medium	1.76	1.19	7.31	2.77	4.68	2.47	0.00	4.75	5.75	10.74	7.60	5.31	21.67	3.75	0.31	3.68
Large	0.00	N.A	N.A	N.A	N.A	N.A	N.A	N.A	N.A	N.A	4.03	N.A	N.A	N.A	N.A	0.54
All class	2.88	2.03	3.45	9.78	3.34	3.37	1.72	1.63	4.26	3.99	4.11	4.24	5.80	3.21	1.80	3.42

Annexure 2i: Percentage of GCA under fruits in Malda District, 1995–1996 and 2015–2016

	Bam.	Ch-I	Ch-II	E.B.	Gaz.	Hbb.	HCP-I	HCP-II	KL-I	KL-II	KL-III	Manik.	Odl M	Rat.-I	Rat.II	District
1995–1996																
Marginal	0.11	0.03	0.15	8.63	0.01	0.00	0.00	0.05	24.64	6.95	0.48	2.94	0.04	0.98	2.73	2.32
Small	0.00	0.03	0.20	13.08	0.21	0.00	0.00	0.09	19.38	14.03	1.73	3.20	0.00	1.29	3.76	2.00
Semi-medium	0.00	0.11	1.47	15.69	0.47	0.00	0.00	0.22	18.69	13.99	1.07	4.58	0.00	2.19	4.11	2.59
Medium	0.00	2.61	3.78	11.58	0.00	0.00	0.00	0.00	0.00	4.67	4.24	2.68	0.00	2.29	7.77	2.58
Large	5.66	N.A	N.A	7.96	5.33	4.57	0.00	0.00	3.70	N.A	4.30	N.A	0.00	0.00	N.A	4.05
All class	0.04	0.16	0.58	11.45	0.19	0.00	0.00	0.09	20.72	9.06	1.48	3.19	0.02	1.62	3.76	2.30

(continued)

2015–2016

	Bam.	Ch-I	Ch-II	E.B.	Gaz.	Hbb.	HCP-I	HCP-II	KL-I	KL-II	KL-III	Manik.	Odl M	Rat.-I	Rat.II	District
Marginal	0.82	0.53	23.60	26.93	5.33	0.00	0.22	0.04	37.33	8.83	0.23	7.84	0.90	5.34	8.62	7.61
Small	0.43	2.21	23.41	34.72	2.99	0.07	0.00	0.08	31.79	9.14	0.27	13.79	0.00	7.92	13.93	8.74
Semi-medium	0.16	0.62	18.16	24.44	1.32	1.26	0.74	2.65	31.46	47.29	0.39	17.84	0.60	10.29	21.96	8.83
Medium	0.00	46.02	13.21	29.35	7.02	0.00	0.00	0.00	17.29	3.88	1.87	35.10	0.00	2.15	2.25	7.14
Large	0.47	N.A	N.A	N.A	N.A	N.A	N.A	N.A	N.A	N.A	0.95	N.A	N.A	N.A	N.A	1.99
All class	0.58	1.59	22.71	28.85	4.12	0.31	0.23	0.44	35.45	14.79	0.29	10.71	0.59	6.53	11.61	8.07

Annexure 2j: Percentage of GCA under total food crops in Malda District, 1995–1996 and 2015–2016

	Bam.	Ch-I	Ch-II	E.B.	Gaz.	Hbb.	HCP-I	HCP-II	KL-I	KL-II	KL-III	Manik.	Odl M	Rat.-I	Rat.II	District
1995–1996																
Marginal	87.87	80.56	84.17	97.85	86.93	96.07	84.50	89.42	67.40	59.81	82.17	91.87	97.22	90.06	97.90	87.25
Small	87.27	79.42	84.15	98.68	88.31	95.83	83.01	89.51	78.89	79.94	85.17	91.32	97.63	90.20	98.16	89.49
Semi-medium	91.37	81.86	83.36	98.07	89.92	93.44	89.49	89.70	83.98	82.28	85.22	92.07	97.96	87.97	98.44	89.99
Medium	90.73	87.82	92.80	98.73	89.37	96.02	77.84	83.07	89.14	95.87	85.52	95.25	97.23	86.58	95.33	90.70
Large	100.00	N.A	N.A	100.00	95.27	100.00	100.00	100.00	100.00	N.A	81.72	N.A	100.00	100.00	N.A	97.80
All class	89.04	80.75	84.48	98.25	88.38	95.41	84.50	89.30	73.90	72.12	84.15	92.07	97.48	88.86	97.91	88.80
2015–2016																
Marginal	72.41	75.19	88.12	81.85	85.18	78.82	75.11	88.17	62.47	58.54	55.68	60.07	94.38	73.00	84.45	76.84
Small	73.89	77.67	89.66	82.26	87.17	77.09	79.54	90.20	65.97	83.07	56.65	65.43	97.68	77.18	84.35	80.00
Semi-medium	73.13	79.72	90.80	85.86	88.28	80.08	82.82	90.90	77.80	89.25	61.62	62.80	97.32	73.63	85.30	82.50
Medium	75.01	91.07	85.81	93.29	92.58	86.20	77.49	71.36	83.47	92.10	66.46	96.12	89.34	79.70	86.51	84.85
Large	84.70	N.A	N.A	N.A	N.A	N.A	N.A	N.A	N.A	N.A	64.79	N.A	N.A	N.A	N.A	85.30
All class	73.06	76.55	88.89	82.87	86.34	79.05	77.02	89.03	64.44	68.63	56.89	62.33	95.56	74.29	84.55	78.60

Annexure 2k: Percentage of GCA under total non-food crops in Malda District, 1995–1996 and 2015–2016

	Bam.	Ch-I	Ch-II	E.B.	Gaz.	Hbb.	HCP-I	HCP-II	KL-I	KL-II	KL-III	Manik.	Odl M	Rat.-I	Rat.II	District
1995–1996																
Marginal	12.13	19.44	15.83	2.15	13.07	3.93	15.50	10.58	32.60	40.19	17.83	8.13	2.78	9.94	2.10	12.75
Small	12.73	20.58	15.85	1.32	11.69	4.17	16.99	10.49	21.11	20.06	14.83	8.68	2.37	9.80	1.84	10.51
Semi-medium	8.63	18.14	16.64	1.93	10.08	6.56	10.51	10.30	16.02	17.72	14.78	7.93	2.04	12.03	1.56	10.01
Medium	9.27	12.18	7.20	1.27	10.63	3.98	22.16	16.93	10.86	4.13	14.48	4.75	2.77	13.42	4.67	9.30
Large	0.00	N. A	N. A	0.00	4.73	0.00	0.00	0.00	0.00	N. A	18.28	N. A	0.00	0.00	N. A	2.20
All class	10.96	19.25	15.52	1.75	11.62	4.59	15.50	10.70	26.10	27.88	15.85	7.93	2.52	11.14	2.09	11.20
2015–2016																
Marginal	27.59	24.81	11.88	18.15	14.82	21.18	24.89	11.82	37.53	41.45	44.32	39.93	5.60	27.00	15.55	23.16
Small	26.12	22.33	10.35	17.74	12.83	22.91	20.46	9.80	34.03	16.93	43.35	34.57	2.29	22.82	15.64	20.00
Semi-medium	26.85	20.28	9.22	14.14	11.72	19.92	17.18	9.10	22.20	10.75	38.38	37.26	2.68	26.40	14.70	17.49
Medium	24.99	8.93	14.19	6.71	7.42	13.83	22.03	28.64	16.53	7.90	33.54	3.88	10.66	20.44	13.49	15.15
Large	15.30	N. A	N. A	N. A	N. A	N. A	N. A	N. A	N. A	N. A	40.41	N. A	N. A	N. A	N. A	15.27
All class	26.94	23.45	11.11	17.13	13.66	20.95	22.97	10.97	35.56	31.37	43.11	37.67	4.44	25.71	15.45	21.40

Source: based on author calculation, data extracted from Agriculture Census, Government of India (data on percentage)

Annexure: Chapter 3

Annexure 3a: Macro soil nutrients

Year	No. of sample	PH			EC				%Of C			Nitrogen			P2 O5			K2O		
		Ac	N	ALK	<1	1–2	2–3	>3	L	M	H	L	M	H	L	M	H	L	M	H
1995–1996	7942	3722	2167	2057	7757	110	63	16	3919	3153	874	5546	2364	32	4358	1967	1620	2432	3443	2070
1996–1997	6758	3212	1553	1993	6587	124	20	28	2833	3134	792	5418	1269	70	3700	1559	1500	2430	2902	1427
1997–1998	6037	2851	1218	1968	5866	110	22	39	2618	2706	713	4428	1579	33	3287	1325	1426	2172	2583	1282
1998–1999	4977	1843	1480	1654	4769	157	27	25	2565	1960	452	3607	1351	19	2915	927	1135	2415	1850	712
1999–2000	5105	1758	2049	1328	4889	160	41	12	2926	1748	428	3305	1792	8	2727	1029	1346	2806	1682	615
2000–2001	5138	1608	2233	1327	4897	199	39	0	2843	1835	457	3457	1564	117	2316	1136	1683	3095	1456	584
2001–2002	4821	1964	1891	997	4612	170	28	8	2550	1759	488	3919	886	17	1925	1110	1782	2742	1377	699
2002–2003	3798	1653	1319	826	3643	125	12	17	1936	1317	358	3059	719	21	1526	886	1153	2026	1139	633
2003–2004	9054	3897	2720	2447	8792	124	145	3	5347	2810	907	5878	3011	164	4893	2268	1903	2537	3716	2811
2004–2005	5873	2642	1822	1409	5757	116	0	0	2760	2468	645	4527	1340	9	3464	1409	1000	1998	2700	1175
2005–2006	8900	4626	1959	2315	8722	90	44	44	3649	4182	1069	6602	2220	81	4718	2224	1958	2761	3914	2225
2006–2007	5502	2369	879	2254	5281	167	15	39	2089	2751	662	3843	1647	12	2917	1044	1541	2531	2091	880
2007–2008	3710	1557	817	1336	3594	74	7	35	2116	1186	408	2834	851	23	2225	706	779	1225	1743	742
2008–2009	5719	1602	2744	1373	5432	229	58	0	3489	1943	287	4413	1292	14	3603	1030	1086	3490	1715	514
2009–2010	5887	2116	2587	1275	5642	177	59	0	3173	2116	589	4522	1346	21	2353	1351	2174	3702	1587	589
2010–2011	3808	1106	1368	1334	3618	190	0	0	1868	1445	495	2925	871	14	991	1027	1790	2092	1066	650
2011–2012	4767	2669	1717	381	4576	144	24	23	2610	1716	381	3662	1090	17	2432	953	1382	2432	1478	857
2012–2013	2819	1184	873	762	2735	42	13	29	1331	790	198	2166	645	10	1155	677	287	1553	873	393
2013–2014	2390	1360	455	575	2344	24	11	11	1625	526	239	1836	546	9	979	575	836	1311	742	337
2014–2015	2643	1128	717	798	2566	51	7	19	1848	498	297	2030	604	9	955	478	1210	1153	903	587
2015–2016	17948	7382	6412	4156	1795	52	1	0	9403	6965	1580	13788	4103	64	9620	4929	6099	7041	8061	2846

Annexure 3b: Macro soil nutrients

Year	No. of sample	S		Zn		Fe		Cu		Mn		B	
		Sufficient	Deficient	Sufficient	Deficient	Sufficient	Deficient	Sufficient	Deficient	Sufficient	Deficient	Sufficient	Deficient
1995–1996	7942	3	7930	278	7664	7511	431	6906	1027	5259	2674	115	7828
1996–1997	6758	21	6722	732	6023	6552	203	5635	886	4367	2150	33	6722
1997–1998	6037	48	5971	360	5678	5544	510	3391	319	2628	1083	59	5979
1998–1999	4977	3	4973	398	4579	4783	194	3821	333	2600	1547	13	4964
1999–2000	5105	161	4943	376	4728	3164	1940	2158	386	1870	674	22	5082
2000–2001	5138	12	5168	247	4936	4889	291	4891	291	3213	1970	25	5158
2001–2002	4821	1	4815	201	4619	4803	17	4411	353	3207	1564	37	4783
2002–2003	3798	2	3791	245	3553	3790	8	3498	284	2621	1161	8	3790
2003–2004	9054	40	9010	662	8388	8565	485	7498	1460	5898	3059	172	8878
2004–2005	5873	0	5873	905	4968	5833	441	4817	4817	441	1517	11	5862
2005–2006	8900	3	8888	675	8225	8363	537	7182	1067	4875	3374	42	8858
2006–2007	5502	70	5431	585	4917	4713	786	3174	348	2382	1104	37	5462
2007–2008	3710	11	3696	172	3505	3325	384	3081	130	2260	951	3	3707
2008–2009	5719	2	5717	954	4765	5685	34	4332	559	3733	1158	26	5693
2009–2010	5887	31	5855	384	5500	5162	717	3648	455	2438	1671	99	5784
2010–2011	3808	20	3787	248	3558	3339	464	2360	294	1577	1081	64	3741
2011–2012	4767	2	4760	167	4600	4508	259	4145	616	3157	1605	69	4698
2012–2013	2819	9	2804	305	2512	2733	85	2350	369	1822	897	14	2804
2013–2014	2390	19	2364	142	2248	2195	202	1342	126	1040	429	23	2367
2014–2015	2643	1	2641	211	2432	2540	103	2029	177	1381	821	7	2636
2015–2016	1794	1	1793	143	1651	1724	70	1377	120	937	557	5	1789

Source: Soil Testing Laboratory Malda, 2018

Annexure 3c: Temperature (in degree Centigrade)

| | Kharif season | | | | | | | | | | | | Rabi season | | | | | | | | Zaid season | | | |
| | Oct | | Nov | | Dec | | Jan | | Feb | | Mar | | June | | July | | Aug | | Sept | | April | | May | |
Year	Max.	Min.	Max.	Min.	Max.	Min.	Max.	Min.	Max.	Min.	Max.	Min.	Max.	Min.	Max.	Min.	Max.	Min.	Max.	Min.	Max.	Min.	Max.	Min.
1995	21.0	15.0	31.0	15.0	27.0	14.0	27.0	9.0	31.0	12.0	40.0	15.0	40.0	18.0	34.0	24.0	25.0	25.0	23.0	23.0	43.0	21.0	44.0	25.0
1996	21.0	15.0	31.0	15.0	25.2	11.4	22.6	11.0	26.5	13.4	32.4	18.6	32.6	25.3	32.3	26.2	31.8	26.0	32.9	26.3	35.7	20.9	36.4	25.2
1997	30.6	21.3	28.2	17.4	22.5	12.4	22.9	9.4	25.8	11.5	31.7	17.5	35.2	25.6	32.0	26.2	32.3	25.8	31.3	24.5	31.4	19.8	35.3	34.0
1998	33.0	23.7	28.8	18.9	25.2	12.6	20.5	9.4	25.9	13.1	29.4	15.4	35.9	26.9	32.4	25.7	31.8	25.9	33.1	25.5	32.2	21.1	34.0	24.1
1999	32.0	21.9	28.6	14.7	25.7	10.7	23.0	9.9	28.4	12.5	33.4	14.2	33.5	23.2	31.5	23.2	31.2	23.0	30.6	22.5	36.2	21.5	33.3	21.6
2000	32.7	20.3	29.3	14.2	25.2	7.9	22.7	6.8	24.0	9.0	29.7	13.2	33.1	21.8	32.6	22.9	32.9	22.9	30.6	21.7	36.3	17.9	32.4	20.2
2001	31.0	23.4	29.2	18.6	23.1	12.1	22.9	8.3	26.4	14.1	32.8	14.2	32.2	25.4	32.6	26.2	33.0	26.1	32.1	25.1	36.0	20.3	32.2	22.4
2002	31.2	21.8	28.7	17.5	24.4	12.8	23.7	11.5	27.2	12.8	31.8	17.2	33.3	25.0	32.6	25.9	32.1	25.7	32.4	25.0	31.6	21.4	32.2	23.4
2003	29.9	23.1	28.4	16.9	24.7	12.8	20.0	8.5	26.0	14.0	29.2	17.2	33.3	25.7	32.6	26.5	32.9	26.7	32.1	25.9	34.2	22.3	34.2	23.4
2004	29.9	21.9	28.3	16.2	25.3	13.6	20.8	10.2	25.2	13.1	33.0	18.7	32.8	25.0	32.0	25.8	32.5	26.2	31.3	25.3	32.9	21.7	35.8	24.2
2005	28.6	22.5	27.6	16.2	25.0	12.3	22.9	10.7	27.5	14.2	31.6	19.0	35.9	25.9	32.0	25.9	32.6	26.6	33.4	25.9	34.5	21.7	33.9	23.7
2006	32.2	23.5	28.1	18.3	25.3	12.9	23.4	10.1	29.5	16.3	33.0	17.5	34.1	26.1	33.1	27.0	33.0	26.5	32.2	25.7	33.7	22.2	34.5	24.5
2007	31.5	23.9	29.2	18.6	24.4	11.9	23.5	9.6	25.6	14.7	30.1	16.7	33.3	25.9	32.0	26.4	32.9	26.9	32.9	26.5	34.4	23.1	36.3	25.1
2008	31.5	23.1	29.0	17.2	24.5	15.4	23.3	11.4	24.4	12.3	31.4	19.1	32.7	25.8	32.0	26.4	32.4	26.6	32.6	25.8	35.6	22.1	35.4	24.6
2009	31.8	22.5	28.8	17.4	24.9	11.5	23.6	12.2	28.6	13.1	32.6	17.6	36.1	26.6	33.2	26.8	32.9	26.5	33.5	26.3	36.3	22.1	23.9	23.8
2010	31.9	23.7	29.3	18.6	24.8	11.4	21.2	9.5	27.9	13.0	34.8	19.4	34.1	26.2	33.4	26.9	33.6	26.9	32.9	26.1	36.9	24.8	35.5	25.0
2011	32.4	23.8	28.7	17.2	23.4	12.6	21.5	8.7	27.7	12.6	32.9	17.8	33.7	26.2	33.1	26.6	32.1	26.3	32.4	26.3	34.0	21.6	34.1	24.2
2012	31.7	22.0	27.5	16.0	21.4	11.1	22.9	11.3	27.6	12.2	33.0	16.7	35.2	27.1	32.9	26.6	33.4	26.9	33.0	26.1	34.6	22.1	37.5	25.6
2013	31.7	22.0	27.5	16.0	21.4	11.1	22.9	11.3	27.6	12.2	33.0	16.7	35.2	27.1	32.9	27.1	33.4	26.9	33.0	26.1	34.6	22.1	37.5	25.6
2014	31.8	23.2	28.8	16.1	22.6	12.3	22.1	11.1	24.6	12.7	31.2	16.8	32.9	26.4	32.9	27.2	32.7	26.6	32.2	25.8	36.8	21.5	37.1	25.2
2015	32.6	23.7	29.8	18.8	24.2	13.2	22.7	11.5	26.6	14.9	31.5	17.6	33.8	26.7	33.0	26.4	32.4	27.0	33.5	26.6	31.9	22.1	34.1	25.0
2016	32.4	24.3	29.5	17.9	24.8	14.2	23.1	10.9	28.3	15.0	32.7	19.6	34.3	27.0	32.4	27.0	33.8	27.2	32.5	26.5	37.1	24.2	33.9	24.7

Annexure 3d: Rainfall (in centimeter)

Year	Kharif			Rabi season							Zaid	
	Oct.	Nov.	Dec.	Jan.	Feb.	Mar.	Jun.	Jul.	Aug.	Sep.	Apr.	May.
1995	8.20	81.44	8.78	3.19	7.29	2.79	300.41	359.58	550.75	716.56	3.45	52.74
1996	60.09	0.00	0.00	9.26	15.80	3.03	199.16	233.80	289.60	391.60	32.98	37.81
1997	25.54	1.41	30.78	16.60	9.47	7.79	206.00	387.56	547.10	266.68	77.92	90.02
1998	232.56	18.27	0.00	5.21	9.39	84.36	173.00	502.27	328.82	316.93	60.61	101.28
1999	176.14	3.69	0.00	0.00	0.08	0.13	348.47	515.93	466.62	656.03	12.09	143.37
2000	58.30	0.00	0.00	0.91	49.60	38.99	324.34	163.14	209.51	560.34	115.43	191.54
2001	295.32	0.00	0.00	0.43	0.00	0.32	356.19	240.38	214.04	246.59	7.70	232.18
2002	44.38	6.33	0.00	8.78	0.42	13.07	206.74	230.72	352.91	391.73	128.47	112.33
2003	332.03	0.00	7.43	3.99	67.54	61.37	325.61	331.26	142.73	234.09	38.43	132.13
2004	515.10	0.00	1.90	24.59	0.00	2.18	332.19	474.73	248.53	159.04	84.71	87.10
2005	193.82	0.00	0.00	12.46	9.81	41.21	87.04	529.21	395.28	187.58	13.78	109.64
2006	66.04	14.02	1.21	0.00	0.00	10.53	169.47	207.03	250.82	381.33	60.86	102.40
2007	47.58	2.02	0.00	0.00	32.61	32.29	388.56	555.98	196.57	235.17	1.67	109.98
2008	62.64	0.00	0.00	51.61	0.11	15.67	375.57	344.90	294.23	211.77	14.82	66.32
2009	209.54	1.62	0.00	0.13	6.91	3.63	67.97	235.80	325.81	109.33	0.00	237.98
2010	102.50	6.04	3.40	0.00	1.03	0.00	262.57	211.42	178.64	218.78	26.50	123.56
2011	24.21	0.00	0.00	1.16	6.93	9.84	320.39	252.63	387.56	248.39	35.59	155.96
2012	83.30	33.42	0.00	11.84	0.00	1.33	118.02	269.41	140.19	233.13	46.41	25.11
2013	221.80	1.40	0.00	0.00	15.00	1.30	230.30	160.30	354.60	144.20	42.80	65.90
2014	19.10	0.00	0.00	0.30	54.40	1.40	263.30	315.10	226.20	267.30	4.20	152.10
2015	17.20	0.00	0.00	18.80	1.60	43.40	257.30	419.10	330.00	193.60	103.70	154.20
2016	72.00	0.00	0.00	16.90	0.00	2.10	157.40	407.30	152.10	337.90	21.60	85.40

Annexure 3e: Humidity (in percentage)

Year	Kharif season								Rabi season												Zaid season			
	Oct		Nov		Dec		Jan		Feb		Mar		June		July		Aug		Sept		April		May	
	Max.	Min.	Max.	Min.	Max.	Min.	Max.	Min.	Max.	Min.	Max.	Min.	Max.	Min.	Max.	Min.	Max.	Min.	Max.	Min.	Max.	Min.	Max.	Min.
1995	93	65	94	45	96	47	97	47	92	38	87	34	92	74	94	74	92	77	94	80	84	37	92	64
1996	95	62	94	47	98	52	95	52	93	37	74	27	89	57	95	77	93	73	94	68	77	27	89	53
1997	95	65	95	59	92	48	97	51	90	42	85	33	91	69	93	73	94	77	94	77	86	43	89	58
1998	95	61	93	53	91	45	97	54	95	49	89	35	93	70	92	70	93	73	94	68	90	47	86	46
1999	87	63	94	59	82	57	95	45	95	37	83	26	93	76	94	83	95	82	94	79	75	28	86	41
2000	69	82	96	50	94	47	98	54	96	45	85	40	97	69	94	78	73	79	94	94	80	36	86	44
2001	93	62	95	59	98	62	99	46	94	41	89	37	88	61	94	83	94	79	95	81	88	52	85	50
2002	91	71	90	61	94	55	99	62	94	44	91	42	91	64	96	80	95	81	92	75	88	57	89	60
2003	93	72	95	53	95	56	96	54	93	41	83	34	93	74	95	80	97	82	95	81	89	26	90	65
2004	93	66	96	58	96	51	98	57	94	51	91	46	94	74	91	75	93	75	91	77	90	32	90	67
2005	96	75	98	62	98	63	95	47	93	39	85	33	92	76	94	78	94	77	96	80	87	40	90	69
2006	95	65	94	55	97	61	97	59	94	42	90	40	93	73	95	77	93	78	93	75	92	62	92	69
2007	95	73	99	53	98	54	99	58	98	52	89	50	92	72	93	76	92	73	94	79	91	55	90	53
2008	95	70	95	53	98	59	98	66	96	45	92	49	93	73	95	79	93	76	95	80	89	57	89	52
2009	96	83	92	74	89	51	98	58	91	47	93	50	90	60	93	76	94	78	92	72	89	53	90	57
2010	89	62	94	59	94	52	95	51	96	49	85	30	88	68	90	71	89	73	92	73	86	53	87	59
2011	93	65	89	52	94	49	97	44	96	53	87	41	85	73	93	78	90	73	92	72	85	47	81	53
2012	91	62	92	48	94	63	93	53	92	44	90	46	90	70	91	77	92	74	92	73	86	42	85	51
2013	92	61	92	52	96	49	95	58	89	33	83	34	88	60	92	73	92	76	92	70	87	34	87	59
2014	91	69	90	56	91	53	91	56	90	44	85	33	92	75	93	72	91	69	91	72	90	45	89	59
2015	90	61	93	56	94	58	91	52	88	41	85	35	92	70	92	73	92	77	89	74	86	46	89	62
2016	89	60	91	56	96	68	93	56	88	33	79	31	89	68	93	74	92	71	94	71	82	45	82	48

Sources: (Annexures 3c–3e) Deputy Director of Agriculture (Admin.), Malda, 2018

Annexure 3f: List of non-physical inputs variables of agricultural development in 1995–1996

Vr.	Bam.	Ch-I	Ch-II	E.B.	Gaz.	Hbb.	HCP-I	HCP-II	KL-I	KL-II	KL-III	Manik.	Odl M	Rat.-I	Rat.II	District
Marginal land size (<1 hectare) class																
X1	69.43	76.63	81.46	74.98	66.77	67.90	84.47	72.67	90.53	84.82	71.29	85.27	74.61	73.51	72.83	75.63
X2	0.52	0.52	0.53	0.54	0.53	0.56	0.54	0.52	0.42	0.48	0.51	0.51	0.52	0.53	0.54	0.52
X3	82.85	86.53	81.40	81.61	75.78	75.27	87.56	88.81	89.71	88.80	91.39	90.18	57.79	92.43	79.62	82.67
X4	2.46	0.88	0.76	3.92	3.41	2.79	0.90	1.25	0.00	0.00	0.15	0.99	7.26	0.88	0.00	1.90
X5	99.98	98.78	97.86	93.10	99.46	98.94	97.72	96.24	97.14	97.01	95.63	96.96	98.41	92.26	95.22	97.22
X6	66.32	57.75	52.36	44.51	80.74	69.36	68.72	75.80	55.43	51.68	62.70	52.76	61.89	54.48	84.59	63.11
X7	33.68	42.25	47.64	55.49	19.26	30.64	31.28	24.20	44.57	48.32	37.30	47.24	38.11	45.52	15.41	36.89
X8	45.43	75.32	82.43	42.17	37.64	25.61	85.82	92.31	19.68	23.70	56.21	28.33	42.15	54.03	69.87	50.46
X9	51.62	8.63	6.78	20.90	48.17	49.47	2.38	0.92	0.54	64.73	5.68	12.57	31.15	2.48	24.51	17.75
X10	0	0	0	0	0	1	0	0	1	1	0	0	1	0	2	0
X11	1	0	0	1	1	5	2	3	1	0	1	1	10	0	1	2
X12	42.32	47.45	78.46	48.72	39.10	32.93	84.11	94.65	31.62	27.97	54.17	28.83	52.69	55.29	55.50	54.28
X13	57.68	52.55	21.54	51.28	60.90	67.07	15.89	5.35	68.38	72.03	45.83	71.17	47.31	44.71	44.50	45.72
X14	11.06	17.36	22.83	32.50	14.99	3.66	15.36	10.31	31.91	10.88	17.00	33.71	31.42	16.38	33.47	20.19
X15	19.11	19.43	20.98	20.63	23.26	21.86	20.48	20.43	20.01	18.92	18.46	21.91	19.49	19.45	18.49	20.19
X16	0.46	0.51	0.73	0.68	1.06	0.86	0.66	0.65	0.59	0.44	0.37	0.87	0.52	0.51	0.37	0.62
X17	0.13	0.14	0.20	0.19	0.30	0.24	0.18	0.18	0.17	0.12	0.10	0.24	0.15	0.14	0.10	0.17
X18	0.01	0.01	0.01	0.01	0.01	0.01	0.01	0.01	0.01	0.01	0.00	0.01	0.01	0.01	0.00	0.01
X19	0.14	0.16	0.22	0.21	0.33	0.26	0.20	0.20	0.18	0.13	0.11	0.27	0.16	0.16	0.12	0.19
X20	0.19	0.20	0.29	0.27	0.42	0.34	0.26	0.26	0.24	0.17	0.15	0.35	0.21	0.20	0.15	0.25
X21	53	85	106	73	124	80	93	97	45	44	49	116	53	73	50	76
X22	13	20	25	18	30	19	22	23	11	10	12	28	13	17	12	18
X23	1	1	2	1	2	1	2	2	1	1	1	2	1	1	1	1
X24	27.52	44.02	54.78	37.84	64.49	41.67	48.44	50.09	23.39	22.63	25.54	59.98	27.52	37.58	26.16	39.44

Vr.	Bam.	Ch-I	Ch-II	E.B.	Gaz.	Hbb.	HCP-I	HCP-II	KL-I	KL-II	KL-III	Manik.	Odl M	Rat.-I	Rat.II	District
X25	8.74	13.99	17.40	12.02	20.49	13.24	15.39	15.91	7.43	7.19	8.11	19.06	8.74	11.94	8.31	12.53
X26	175	280	348	240	310	265	308	318	149	144	162	321	175	239	166	240
X27	101	162	202	139	237	153	178	184	186	197	206	221	101	138	196	174
X28	18.02	14.06	19.60	16.79	16.45	20.04	13.38	10.12	15.70	15.85	17.50	19.83	19.47	21.40	17.34	17.04
X29	81.98	85.94	80.40	83.21	83.55	79.96	86.62	89.88	84.30	84.15	82.50	80.17	80.53	78.60	82.66	82.96
X30	32.06	35.20	50.49	47.06	53.11	39.24	45.59	45.06	40.91	30.12	25.56	39.73	35.78	35.39	25.88	40.15
X31	45	45	46	45	45	45	45	44	45	43	45	46	44	46	45	45
X32	6	6	6	5	6	6	7	7	7	6	7	7	6	6	6	6
Under small land size (1–2 hectare) class																
X1	17.74	17.15	13.20	16.44	22.19	22.63	9.46	21.13	5.69	9.47	16.65	10.05	18.16	12.40	18.49	16.16
X2	1.56	1.64	1.62	1.59	1.60	1.64	1.61	1.61	1.64	1.60	1.66	1.66	1.55	1.64	1.64	1.62
X3	81.84	86.11	82.03	65.33	70.54	69.30	86.37	88.19	81.18	82.10	87.12	88.80	48.60	82.85	72.27	75.83
X4	0.00	0.00	0.00	6.00	1.44	0.87	0.00	0.00	1.69	0.00	0.00	0.62	3.41	0.00	0.00	1.11
X5	99.90	98.88	98.43	87.49	98.74	99.02	97.42	97.79	96.68	98.66	88.06	94.89	97.80	91.48	93.11	96.52
X6	58.87	38.35	36.66	21.57	63.54	57.68	44.93	73.00	27.79	17.75	37.73	22.28	43.61	46.64	66.41	49.57
X7	41.13	61.65	63.34	78.43	36.46	42.32	55.07	27.00	72.21	82.25	62.27	77.72	56.39	53.36	33.59	50.43
X8	50.54	76.28	80.76	46.07	35.57	27.66	84.04	93.71	30.36	26.10	50.18	30.17	44.84	50.77	67.03	51.22
X9	57.89	5.65	5.51	21.92	44.21	59.15	0.22	2.58	2.89	24.82	1.88	9.26	22.73	2.12	25.38	20.15
X10	0	8	6	21	2	39	10	8	33	55	3	7	30	2	83	17
X11	1	1	0	0	0	3	1	1	0	6	1	1	1	0	1	1
X12	43.11	51.31	85.31	50.88	37.34	34.16	80.42	93.22	43.27	34.05	54.12	34.47	55.50	54.92	63.82	55.38
X13	56.89	48.69	14.69	49.12	62.66	65.84	19.58	6.78	56.73	65.95	45.88	65.53	44.50	45.08	36.18	44.62
X14	18.66	18.52	18.66	19.61	25.86	23.83	17.28	20.86	16.15	16.50	17.67	18.15	18.89	17.67	17.94	19.08
X15	19.98	19.82	19.98	21.05	24.05	25.77	18.43	22.44	17.16	17.56	18.87	19.40	20.24	18.87	19.17	20.19
X16	1.01	0.98	1.01	1.28	3.01	2.45	0.63	1.62	0.32	0.42	0.74	0.87	1.08	0.74	0.81	1.13
X17	0.27	0.26	0.27	0.34	0.80	0.65	0.17	0.43	0.08	0.11	0.20	0.23	0.29	0.20	0.22	0.30

Vr.	Bam.	Ch-I	Ch-II	E.B.	Gaz.	Hbb.	HCP-I	HCP-II	KL-I	KL-II	KL-III	Manik.	Odl M	Rat.-I	Rat.II	District
X18	0.05	0.05	0.05	0.06	0.14	0.11	0.03	0.08	0.01	0.02	0.03	0.04	0.05	0.03	0.04	0.05
X19	0.12	0.11	0.12	0.15	0.35	0.28	0.07	0.19	0.04	0.05	0.09	0.10	0.12	0.09	0.09	0.13
X20	0.58	0.56	0.58	0.73	1.72	1.40	0.36	0.93	0.18	0.24	0.42	0.50	0.62	0.42	0.47	0.65
X21	28	40	36	32	81	54	21	61	8	10	21	27	26	26	27	33
X22	10	14	13	11	29	19	8	22	3	4	7	10	9	9	10	12
X23	1	2	2	2	4	3	1	3	0	0	1	1	1	1	1	2
X24	32.90	46.38	42.07	37.00	73.52	62.59	24.69	70.92	8.86	11.87	23.91	31.42	30.73	29.69	31.91	37.23
X25	15.59	21.98	19.93	17.53	44.30	29.65	11.70	33.60	4.20	5.62	11.33	14.89	14.56	14.07	15.12	18.27
X26	203	286	260	228	377	386	152	238	155	73	148	194	190	183	197	218
X27	118	167	151	133	236	225	189	202	139	149	199	113	110	107	115	157
X28	92.84	94.85	93.15	92.33	92.89	92.61	92.17	96.17	92.68	92.10	92.30	93.06	92.54	92.77	92.88	93.02
X29	7.16	5.15	6.85	7.67	7.11	7.39	7.83	3.83	7.32	7.90	7.70	6.94	7.46	7.23	7.12	6.98
X30	46.77	44.99	46.70	58.91	58.73	42.74	29.15	54.82	34.68	29.20	34.08	40.19	49.72	34.08	37.51	48.89
X31	47	48	48	48	48	47	47	47	48	46	46	48	48	49	47	47
X32	6	5	6	5	7	7	7	7	7	6	7	7	6	6	6	6
Under semi-medium land size (2–4 hectares) class																
X1	10.43	5.41	4.63	6.16	9.13	7.70	4.96	5.52	2.99	4.28	9.35	3.46	5.48	10.00	7.20	6.55
X2	2.72	2.74	2.74	2.74	2.73	2.71	2.74	2.75	2.82	2.70	2.75	2.75	2.73	2.75	2.76	2.74
X3	83.06	71.41	68.73	70.94	67.94	66.61	80.20	84.95	81.11	76.96	87.22	80.61	61.20	81.94	80.34	74.91
X4	0.00	0.00	0.00	0.00	0.67	0.92	0.00	0.00	0.00	0.00	0.00	2.09	1.04	0.00	0.00	0.37
X5	100.00	98.13	97.52	84.29	99.13	98.80	96.60	94.64	98.17	99.06	92.30	92.07	97.76	92.15	95.78	96.02
X6	44.73	22.24	38.08	11.20	52.56	59.83	49.90	62.52	16.36	43.22	40.71	29.51	21.39	45.92	56.76	43.50
X7	55.27	77.76	61.92	88.80	47.44	40.17	50.10	37.48	83.64	56.78	59.29	70.49	78.61	54.08	43.24	56.50
X8	57.72	72.93	75.95	40.63	37.31	28.25	63.25	83.85	23.06	44.21	49.19	33.19	43.39	53.99	78.77	49.98
X9	47.30	8.70	2.17	15.54	38.68	72.75	0.41	0.24	3.29	50.66	4.07	18.02	33.41	6.33	17.99	21.80
X10	0	0	0	0	0	1	0	0	0	0	0	0	6	0	1	0

(continued)

Vr.	Bam.	Ch-I	Ch-II	E.B.	Gaz.	Hbb.	HCP-I	HCP-II	KL-I	KL-II	KL-III	Manik.	OdI M	Rat.-I	Rat.II	District
X11	1	0	1	0	0	0	1	1	1	0	1	1	0	0	0	1
X12	52.46	51.75	84.34	49.23	39.60	34.86	78.07	91.62	31.95	49.80	49.08	34.54	52.11	53.61	66.30	53.83
X13	47.54	48.25	15.66	50.77	60.40	65.14	21.93	8.38	68.05	50.20	50.92	65.46	47.89	46.39	33.70	46.17
X14	23.50	20.84	21.28	22.41	29.41	25.67	21.06	21.91	19.54	19.74	21.83	20.77	21.00	23.50	20.92	22.22
X15	23.94	20.04	20.68	22.35	32.60	27.12	20.36	21.61	18.14	18.42	21.49	19.94	20.28	23.94	20.16	22.07
X16	1.34	0.69	0.80	1.08	2.79	1.87	0.75	0.95	0.38	0.42	0.93	0.68	0.73	1.34	0.71	1.03
X17	0.37	0.19	0.22	0.30	0.77	0.52	0.21	0.26	0.10	0.12	0.26	0.19	0.20	0.37	0.20	0.29
X18	0.10	0.05	0.06	0.08	0.21	0.14	0.06	0.07	0.03	0.03	0.07	0.05	0.06	0.10	0.05	0.08
X19	0.22	0.11	0.13	0.17	0.45	0.30	0.12	0.15	0.06	0.07	0.15	0.11	0.12	0.22	0.12	0.17
X20	0.65	0.34	0.39	0.52	1.35	0.91	0.36	0.46	0.18	0.20	0.45	0.33	0.35	0.65	0.34	0.50
X21	24	18	19	18	52	30	15	23	7	7	19	14	12	33	18	21
X22	9	7	7	6	19	11	5	8	3	3	7	5	4	12	6	7
X23	2	1	1	1	3	2	1	2	0	0	1	1	1	2	1	1
X24	38.48	29.52	31.47	28.62	84.23	49.19	24.29	37.50	11.71	11.58	31.27	23.34	19.15	54.29	28.65	33.55
X25	20.79	15.95	17.00	15.46	45.51	26.58	13.12	20.26	6.33	6.26	16.89	12.61	10.35	29.33	15.48	18.13
X26	213	164	174	159	367	273	135	208	165	64	173	129	106	301	159	186
X27	156	120	128	116	242	200	109	152	144	156	127	199	206	220	116	159
X28	95.56	93.56	94.77	97.22	94.60	94.40	96.11	95.45	94.46	96.07	94.20	94.71	94.36	94.51	94.59	94.97
X29	4.44	6.44	5.23	2.78	5.40	5.60	3.89	4.55	5.54	3.93	5.80	5.29	5.64	5.49	5.41	5.03
X30	69.21	35.73	41.25	55.58	48.60	46.52	38.49	49.19	19.40	41.85	48.15	34.86	37.80	59.21	36.76	48.62
X31	51	51	51	51	51	51	51	51	51	51	51	51	51	51	51	51
X32	5	5	5	5	6	6	7	7	7	6	7	7	6	6	6	6
Under medium land size (4–10 hectares) class																
X1	2.34	0.81	0.72	2.37	1.88	1.72	1.09	0.64	0.77	1.43	2.65	1.22	1.73	4.04	1.49	1.64
X2	5.49	5.70	5.47	5.51	5.48	5.48	5.48	5.59	5.73	5.37	5.42	5.43	5.50	5.41	5.41	5.48
X3	57.89	56.00	65.51	31.14	59.39	49.85	0.00	50.70	0.00	35.10	72.80	47.04	13.65	80.53	65.66	50.68

Vr.	Bam.	Ch-I	Ch-II	E.B.	Gaz.	Hbb.	HCP-I	HCP-II	KL-I	KL-II	KL-III	Manik.	Odl M	Rat.-I	Rat.II	District
X4	0.00	0.00	0.00	0.00	0.00	0.00	0.00	0.00	0.00	0.00	6.31	0.00	0.00	0.00	0.00	0.43
X5	99.72	96.83	98.32	89.84	97.57	95.85	97.99	94.58	99.76	99.91	95.42	90.81	99.16	85.35	88.58	94.30
X6	32.35	33.26	24.01	19.63	37.22	38.88	100.00	78.44	34.55	2.82	24.25	27.73	12.54	49.37	16.71	35.13
X7	67.65	66.74	75.99	80.37	62.78	61.12	0.00	21.56	65.45	97.18	75.75	72.27	87.46	50.63	83.29	64.87
X8	35.26	53.28	66.70	40.95	33.89	30.54	32.64	66.41	38.79	64.90	55.00	31.67	28.98	48.05	49.62	41.67
X9	73.48	9.98	3.41	14.31	37.85	70.59	3.20	2.12	16.25	1.22	6.71	15.08	71.64	2.28	19.80	23.25
X10	0	0	0	0	0	0	0	0	0	2	0	0	0	0	0	0
X11	0	1	0	0	0	0	1	1	1	0	1	2	0	1	0	0
X12	38.65	49.19	80.38	49.42	33.88	36.62	61.93	74.93	46.18	49.29	49.29	41.61	32.53	56.88	52.50	47.86
X13	61.35	50.81	19.62	50.58	66.12	63.38	38.07	25.07	53.82	50.71	50.71	58.39	67.47	43.12	47.50	52.14
X14	27.99	24.72	25.05	29.88	32.49	29.92	25.72	24.83	24.60	25.34	27.38	26.94	26.83	31.99	25.44	27.27
X15	28.82	20.34	21.20	23.72	30.48	33.83	22.93	20.62	20.02	21.95	27.24	26.09	25.80	39.18	22.21	25.63
X16	1.45	0.50	0.60	2.00	2.76	2.01	0.79	0.53	0.46	0.68	1.27	1.15	1.11	2.61	0.71	1.24
X17	0.29	0.10	0.12	0.39	0.54	0.40	0.16	0.11	0.09	0.13	0.25	0.23	0.22	0.52	0.14	0.25
X18	0.27	0.09	0.11	0.37	0.51	0.37	0.15	0.10	0.09	0.13	0.23	0.21	0.20	0.48	0.13	0.23
X19	0.08	0.03	0.03	0.10	0.14	0.10	0.04	0.03	0.02	0.04	0.07	0.06	0.06	0.14	0.04	0.06
X20	0.82	0.28	0.34	1.13	1.56	1.14	0.45	0.30	0.26	0.38	0.72	0.65	0.63	1.47	0.40	0.70
X21	14	6	8	16	26	16	9	6	4	8	14	12	9	29	7	12
X22	5	2	3	6	10	6	4	2	2	3	5	5	3	11	3	5
X23	1	1	1	1	2	1	1	1	0	1	1	1	1	3	1	1
X24	44.36	19.42	23.74	48.75	81.75	50.83	29.12	19.13	13.40	25.72	44.49	37.83	28.32	89.56	22.49	38.59
X25	25.82	11.30	13.81	28.37	47.58	29.58	16.94	11.13	7.80	14.97	25.89	22.02	16.48	52.12	13.09	22.46
X26	211	92	113	132	388	241	138	191	264	122	211	180	135	425	107	197
X27	154	168	187	170	284	177	101	177	149	169	155	132	198	211	186	175
X28	95.15	97.74	94.29	94.31	97.46	96.61	98.54	98.51	96.23	97.08	97.48	94.46	96.43	94.73	98.36	96.49
X29	4.85	2.26	5.71	5.69	2.54	3.39	1.46	1.49	3.77	2.92	2.52	5.54	3.57	5.27	1.64	3.51

(continued)

Vr.	Bam.	Ch-I	Ch-II	E.B.	Gaz.	Hbb.	HCP-I	HCP-II	KL-I	KL-II	KL-III	Manik.	Odl M	Rat.-I	Rat.II	District
X30	42.44	44.62	37.45	58.47	60.64	58.85	23.11	35.56	33.58	39.90	37.25	33.48	32.54	56.39	20.75	60.84
X31	50	51	50	51	50	50	50	50	51	50	50	51	51	51	50	50
X32	5	5	5	7	7	6	6	6	6	6	6	6	6	6	6	6
Under large land size(>10 hectares) class																
X1	0.05	N.A	N.A	0.04	0.03	0.05	0.02	0.04	0.03	N.A	0.07	N.A	0.03	0.05	N.A	0.03
X2	12.30	N.A	N.A	18.27	10.40	10.44	10.40	10.40	10.40	N.A	10.40	N.A	10.40	10.40	N.A	11.27
X3	0.00	N.A	N.A	100.00	67.31	0.00	100.00	100.00	0.00	N.A	0.00	N.A	100.00	100.00	N.A	55.38
X4	0.00	N.A	N.A	0.00	0.00	0.00	0.00	0.00	0.00	N.A	0.00	N.A	0.00	0.00	N.A	0.00
X5	99.03	N.A	N.A	92.04	96.15	100.00	100.00	99.04	100.00	N.A	43.27	N.A	100.00	91.35	N.A	91.76
X6	0.00	N.A	N.A	0.00	66.03	0.00	100.00	100.00	0.00	N.A	0.00	N.A	100.00	0.00	N.A	27.87
X7	100.00	N.A	N.A	100.00	33.97	100.00	0.00	0.00	100.00	N.A	100.00	N.A	0.00	100.00	N.A	72.13
X8	16.50	N.A	N.A	39.30	26.28	13.30	0.00	11.54	13.46	N.A	9.62	N.A	17.31	14.42	N.A	19.27
X9	58.82	N.A	N.A	0.00	29.27	44.00	0.00	50.00	42.86	N.A	0.00	N.A	55.56	60.00	N.A	26.05
X10	0	N.A	N.A	1	0	1	0	0	0	N.A	0	N.A	0	0	N.A	0
X11	0	N.A	N.A	0	1	0	0	0	0	N.A	1	N.A	0	0	N.A	0
X12	45.09	N.A	N.A	39.30	43.20	12.69	70.00	10.43	12.96	N.A	45.05	N.A	18.87	15.79	N.A	22.11
X13	84.91	N.A	N.A	60.70	56.80	87.31	30.00	89.57	87.04	N.A	84.95	N.A	81.13	84.21	N.A	77.89
X14	36.87	N.A	N.A	0.00	0.00	0.00	0.00	0.00	0.00	N.A	41.25	N.A	0.00	0.00	N.A	7.81
X15	14.00	N.A	N.A	21.00	27.21	29.32	26.32	27.35	27.11	N.A	29.36	N.A	27.32	29.32	N.A	25.83
X16	0.00	N.A	N.A	0.04	0.03	0.05	0.02	0.04	0.03	N.A	0.00	N.A	0.03	0.05	N.A	0.03
X17	0.00	N.A	N.A	0.00	0.00	0.00	0.00	0.00	0.00	N.A	0.00	N.A	0.00	0.00	N.A	0.00
X18	0.00	N.A	N.A	0.00	0.00	0.00	0.00	0.00	0.00	N.A	0.00	N.A	0.00	0.00	N.A	0.00
X19	0.00	N.A	N.A	0.00	0.00	0.00	0.00	0.00	0.00	N.A	0.00	N.A	0.00	0.00	N.A	0.00
X20	0.00	N.A	N.A	0.00	0.00	0.00	0.00	0.00	0.00	N.A	0.00	N.A	0.00	0.00	N.A	0.00
X21	0	N.A	N.A	0	0	0	0	0	0	N.A	0	N.A	0	0	N.A	0
X22	0	N.A	N.A	0	0	0	0	0	0	N.A	0	N.A	0	0	N.A	0

Vr.	Bam.	Ch-I	Ch-II	E.B.	Gaz.	Hbb.	HCP-I	HCP-II	KL-I	KL-II	KL-III	Manik.	Odl M	Rat.-I	Rat.II	District
X23	0	N.A	N.A	0	0	0	0.00	0	0	N.A	0	N.A	0	0	N.A	0
X24	0.00	N.A	N.A	0.00	0.00	0.00	0.00	0.00	0.00	N.A	0.00	N.A	0.00	0.00	N.A	0.00
X25	0.00	N.A	N.A	0.00	0.00	0.00	0.00	0.00	0.00	N.A	0.00	N.A	0.00	0.00	N.A	0.00
X26	204	N.A	N.A	153	426	306	307	303	102	N.A	245	N.A	325	217	N.A	259
X27	201	N.A	N.A	232	189	209	201	225	251	N.A	251	N.A	162	193	N.A	211
X28	98.83	N.A	N.A	99.84	98.90	98.87	0.00	0.00	99.83	N.A	98.84	N.A	0.00	98.84	N.A	46.26
X29	1.17	N.A	N.A	0.16	1.10	1.13	1.11	1.14	0.17	N.A	1.16	N.A	1.15	1.16	N.A	0.94
X30	50.51	N.A	N.A	55.56	55.76	50.91	25.25	50.51	25.25	N.A	50.51	N.A	25.25	50.51	N.A	44.00
X31	50	N.A	N.A	0	0	0	0	0	0	N.A	51	N.A	0	0	N.A	51
X32	6	N.A	N.A	0	0	0	0	0	0	N.A	6	N.A	0	0	N.A	7
Under all land size classes																
X1	100.00	100.00	100.00	100.00	100.00	100.00	100.00	100.00	100.00	100.00	100.00	100.00	100.00	100.00	100.00	100.00
X2	1.01	0.79	0.73	0.89	1.00	0.98	0.72	0.83	0.56	0.74	0.97	0.70	0.86	1.01	0.87	0.86
X3	79.03	81.76	78.50	67.77	70.36	68.49	78.66	86.28	79.71	80.05	85.78	84.52	50.72	85.27	76.18	75.65
X4	0.79	0.38	0.38	3.14	1.70	1.44	0.47	0.49	0.25	0.00	0.98	0.97	4.13	0.28	0.00	1.20
X5	99.93	98.57	97.98	89.37	98.93	98.65	97.48	96.48	97.44	97.95	92.40	95.27	98.19	90.58	94.07	96.44
X6	53.62	43.33	44.06	27.73	63.45	59.93	63.16	72.53	43.53	38.75	44.32	40.73	43.73	49.32	66.29	52.09
X7	46.38	56.67	55.94	72.27	36.55	40.07	36.84	27.47	56.47	61.25	55.68	59.27	56.27	50.68	33.71	47.91
X8	48.79	73.89	79.99	42.75	36.44	27.30	76.46	89.79	23.25	31.68	52.29	29.70	41.69	51.92	69.13	49.54
X9	53.93	7.70	5.52	19.13	43.42	60.31	1.63	1.53	3.57	41.77	4.50	12.93	31.82	3.55	22.78	19.81
X10	0	0	0	0	0	1	0	0	0	1	0	0	0	0	1	0
X11	1	1	0	1	0	3	1	2	1	0	1	1	4	0	1	1
X12	44.66	49.56	86.48	49.45	38.14	34.05	80.48	92.63	34.60	35.44	51.99	31.86	51.40	54.95	60.46	53.81
X13	55.34	50.44	13.52	50.55	61.86	65.95	19.52	7.37	65.40	64.56	48.01	68.14	48.60	45.05	39.54	46.19
X14	11.05	18.03	17.19	19.25	20.64	9.02	16.61	11.19	25.97	29.27	15.30	27.78	26.18	16.19	31.27	19.66
X15	19.62	19.60	20.91	20.97	24.76	22.96	20.26	20.91	19.54	18.76	18.79	21.56	19.77	19.78	18.76	20.46

(continued)

Vr.	Bam.	Ch-I	Ch-II	E.B.	Gaz.	Hbb.	HCP-I	HCP-II	KL-I	KL-II	KL-III	Manik.	Odl M	Rat.-I	Rat.II	District
X16	0.61	0.60	0.81	0.82	1.44	1.14	0.71	0.81	0.59	0.47	0.47	0.92	0.63	0.63	0.47	0.74
X17	0.17	0.17	0.22	0.23	0.39	0.31	0.19	0.22	0.16	0.13	0.13	0.25	0.17	0.17	0.13	0.20
X18	0.02	0.02	0.03	0.03	0.05	0.04	0.02	0.03	0.02	0.01	0.01	0.03	0.02	0.02	0.01	0.02
X19	0.14	0.14	0.19	0.20	0.34	0.27	0.17	0.19	0.14	0.11	0.11	0.22	0.15	0.15	0.11	0.18
X20	0.28	0.27	0.37	0.37	0.65	0.52	0.32	0.37	0.27	0.21	0.21	0.42	0.29	0.29	0.21	0.34
X21	40	47	52	45	96	61	42	61	19	21	34	50	32	54	34	46
X22	11	13	15	12	27	17	12	17	5	6	10	14	9	15	9	13
X23	1	1	2	1	3	2	1	2	1	1	1	1	1	2	1	1
X24	33.08	38.84	42.73	36.58	78.56	50.09	34.32	50.00	15.68	17.29	28.14	41.29	26.51	43.99	27.85	37.66
X25	15.34	18.01	19.82	16.96	36.43	23.23	15.92	23.19	7.27	8.02	13.05	19.15	12.29	20.40	12.92	17.47
X26	169	199	219	187	402	257	176	256	80	89	244	212	136	225	143	200
X27	123	161	198	136	241	161	183	186	180	191	200	218	116	146	188	175
X28	37.33	25.32	30.11	13.35	33.75	32.67	25.06	31.86	3.62	14.96	16.73	17.94	26.04	28.52	24.69	24.13
X29	62.67	74.68	69.89	86.65	66.25	67.33	74.94	68.14	96.38	85.04	83.27	82.06	73.96	71.48	75.31	75.87
X30	35.63	35.45	47.83	48.43	54.49	47.32	41.65	47.85	34.87	27.40	27.67	44.05	37.01	37.16	27.42	41.78
X31	46	46	47	45	46	46	44	46	46	44	45	46	44	46	46	46
X32	5	5	6	5	6	6	7	7	7	6	7	7	6	6	6	6

Annexure 3g: List of non-physical input variables development of agricultural in 2015–2016

Vr.	Bam.	Ch-II	E.B.	Gaz.	Hbb.	HCP-I	HCP-II	KL-I	KL-II	KL-III	Manik.	Odl M	Rat.-I	Rat.II	District
Marginal land size (<1 hectare) class															
X1	81.25	85.78	82.24	79.53	82.03	77.48	85.38	84.03	90.65	85.68	87.47	83.65	83.54	84.94	83.99	83.40
X2	0.47	0.44	0.44	0.44	0.46	0.48	0.45	0.44	0.38	0.47	0.44	0.46	0.46	0.44	0.44	0.45
X3	72.24	83.17	22.72	83.50	76.70	42.97	68.06	41.25	98.02	78.20	61.68	82.00	51.05	87.93	44.65	66.47
X4	3.48	0.38	0.04	0.20	0.17	8.89	7.22	2.32	0.00	9.07	5.22	0.05	7.66	0.50	0.00	2.72
X5	96.71	98.17	96.13	78.56	98.54	95.02	99.88	97.84	97.26	99.48	97.53	91.79	98.17	94.79	99.51	95.69

Vr.	Bam.	Ch-I	Ch-II	E.B.	Gaz.	Hbb.	HCP-I	HCP-II	KL-I	KL-II	KL-III	Manik.	Odl M	Rat.-I	Rat.II	District
X6	79.51	80.35	73.58	90.90	98.89	82.03	64.93	86.53	93.60	74.11	67.55	79.32	74.73	73.78	98.56	82.54
X7	20.49	19.65	26.42	9.10	1.11	17.97	35.07	13.47	6.40	25.89	32.45	20.68	25.27	26.22	1.44	17.46
X8	54.05	89.02	91.62	32.51	53.21	37.75	98.06	94.59	42.76	36.31	87.59	57.29	43.21	76.08	81.02	63.98
X9	58.07	1.82	54.77	86.82	59.04	47.33	1.37	0.72	18.29	14.25	2.14	4.13	30.98	7.09	4.72	23.77
X10	0	1	5	31	0	1	4	0	2	1	1	1	17	4	5	4
X11	6	0	8	6	3	6	11	3	16	7	4	4	16	10	15	9
X12	53.24	62.16	80.31	50.86	41.75	35.29	92.39	95.77	46.59	38.54	71.12	59.63	43.99	83.07	81.08	62.44
X13	46.76	37.84	19.69	49.14	58.25	64.71	7.61	4.23	53.41	61.46	28.88	40.37	56.01	16.93	18.92	37.56
X14	54.66	64.99	77.37	75.14	47.96	46.86	54.16	51.99	75.48	54.34	62.64	70.84	72.44	76.09	74.17	63.94
X15	52.86	55.18	57.87	76.23	66.16	48.43	49.35	45.23	68.68	50.61	38.61	64.15	46.82	42.98	59.38	54.84
X16	1.30	1.39	1.49	1.43	2.22	1.91	1.16	1.39	1.53	1.21	0.73	1.35	0.68	1.70	1.16	1.38
X17	0.13	0.14	0.16	0.15	0.23	0.20	0.12	0.14	0.16	0.13	0.08	0.14	0.07	0.18	0.12	0.14
X18	0.04	0.04	0.04	0.04	0.06	0.05	0.03	0.04	0.04	0.03	0.02	0.04	0.02	0.05	0.03	0.04
X19	0.87	0.93	1.00	0.96	1.48	1.28	0.78	0.93	1.02	0.81	0.49	0.90	0.46	1.14	0.78	0.92
X20	0.26	0.27	0.30	0.28	0.44	0.38	0.23	0.28	0.30	0.24	0.15	0.27	0.13	0.34	0.23	0.27
X21	41	55	59	34	71	54	55	54	27	26	28	41	15	57	43	44
X22	18	25	26	15	32	24	25	24	12	11	13	18	7	26	19	20
X23	18	24	25	15	31	24	24	24	12	11	12	18	7	25	19	19
X24	51.11	68.67	73.42	42.53	88.50	68.30	68.53	67.89	74.06	62.11	35.32	51.63	69.09	71.59	53.40	63.08
X25	61.24	71.97	74.88	76.00	84.09	71.75	71.88	71.50	80.82	79.63	71.59	81.56	71.67	73.76	62.64	73.67
X26	304	408	436	253	526	406	407	403	202	191	210	397	213	425	317	340
X27	230	309	331	192	299	246	257	266	253	276	299	268	210	219	240	260
X28	20.97	21.02	26.73	6.56	34.94	36.87	33.07	31.25	17.63	21.65	22.03	18.63	29.38	23.25	23.44	24.50
X29	79.03	78.98	73.27	93.44	65.06	63.13	66.93	68.75	82.37	78.35	77.97	81.37	70.62	76.75	76.56	75.50
X30	54.09	57.91	62.35	59.64	62.45	59.72	48.32	58.00	59.68	50.40	60.64	56.21	48.35	60.76	48.37	57.39
X31	46	47	47	47	47	46	46	46	47	46	46	47	47	47	46	46

(continued)

Vr.	Bam.	Ch-I	Ch-II	E.B.	Gaz.	Hbb.	HCP-I	HCP-II	KL-I	KL-II	KL-III	Manik.	Odl M	Rat.-I	Rat.II	District
X32	5	5	5	5	6	5	7	6	7	6	6	5	5	6	6	6
Under small land size (1–2 hectare) class																
X1	14.22	10.60	13.83	14.86	13.67	16.12	10.95	11.74	7.86	9.92	9.16	14.80	13.73	11.96	11.93	12.65
X2	1.51	1.49	1.50	1.49	1.51	1.52	1.49	1.49	1.49	1.49	1.49	1.49	1.50	1.49	1.49	1.50
X3	60.86	40.64	46.82	77.81	72.99	48.93	37.72	31.67	99.06	97.74	76.89	80.43	46.86	82.44	39.64	62.19
X4	0.00	0.00	0.00	0.30	0.07	1.74	0.00	0.00	0.08	0.00	0.99	0.03	4.26	0.00	0.00	0.44
X5	97.46	97.09	98.08	76.06	98.48	95.53	99.22	98.59	99.25	99.93	98.84	95.47	95.35	94.46	99.41	95.60
X6	55.27	65.71	77.74	80.62	97.78	70.53	46.14	68.32	96.39	36.38	35.45	34.63	74.29	49.94	91.71	68.27
X7	44.73	34.29	22.26	19.38	2.22	29.47	53.86	31.68	3.61	63.62	64.55	65.37	25.71	50.06	8.29	31.73
X8	56.46	87.75	95.18	31.44	47.78	44.19	96.88	96.57	28.93	64.18	89.52	62.49	47.06	77.51	73.76	63.72
X9	66.87	0.74	56.52	84.26	65.72	46.07	0.09	0.41	0.14	1.24	1.38	6.21	24.41	1.69	12.53	26.48
X10	1	8	15	22	2	39	14	8	33	55	3	7	57	2	84	19
X11	3	1	8	4	0	5	8	1	1	69	9	3	6	0	10	7
X12	41.62	63.26	84.79	50.19	38.28	41.37	93.79	73.96	39.00	64.58	69.84	62.46	45.30	82.93	76.74	62.08
X13	58.38	36.74	15.21	49.81	61.72	58.63	6.21	26.04	61.00	35.42	30.16	37.54	54.70	17.07	23.26	37.92
X14	66.42	71.86	78.12	77.56	80.45	81.26	74.21	75.50	67.74	73.96	62.18	76.75	66.16	68.76	71.66	72.84
X15	53.78	45.54	57.42	59.77	74.98	79.19	42.12	48.92	59.69	40.83	41.45	55.49	36.62	55.56	44.51	53.06
X16	2.55	1.13	1.66	1.76	2.44	2.62	0.98	1.28	0.87	0.92	0.51	1.57	0.74	1.58	1.09	1.45
X17	0.41	0.31	0.45	0.48	0.66	0.71	0.27	0.35	0.24	0.25	0.14	0.43	0.20	0.43	0.29	0.37
X18	0.16	0.12	0.18	0.19	0.26	0.28	0.10	0.14	0.09	0.10	0.05	0.17	0.08	0.17	0.12	0.15
X19	1.69	1.28	1.87	1.99	2.75	2.96	1.11	1.45	0.99	1.04	0.57	1.78	0.83	1.78	1.23	1.56
X20	0.42	0.32	0.47	0.50	0.69	0.74	0.28	0.36	0.25	0.26	0.14	0.44	0.21	0.45	0.31	0.39
X21	16	16	23	14	26	26	15	17	8	7	7	18	6	19	14	15
X22	7	7	11	6	12	12	7	8	3	3	3	8	3	9	6	7
X23	8	8	12	7	13	13	8	9	4	3	4	9	3	10	7	8
X24	59.87	57.90	87.16	50.80	95.59	98.15	57.00	64.41	78.67	64.35	26.83	67.77	71.02	70.62	50.71	66.72

Vr.	Bam.	Ch-I	Ch-II	E.B.	Gaz.	Hbb.	HCP-I	HCP-II	KL-I	KL-II	KL-III	Manik.	Odl M	Rat.-I	Rat.II	District
X25	54.81	54.00	66.13	71.06	69.62	70.68	53.63	56.70	81.88	80.09	71.12	81.09	78.71	59.27	51.02	66.65
X26	266	358	388	246	425	437	254	287	228	108	219	301	294	314	226	290
X27	196	190	286	166	313	241	290	250	194	281	288	222	211	231	266	242
X28	91.47	88.96	89.30	95.50	77.54	87.11	84.86	78.66	91.61	90.10	87.04	92.21	81.05	88.30	84.98	87.25
X29	8.53	11.04	10.70	4.50	22.46	12.89	15.14	21.34	8.39	9.90	12.96	7.79	18.95	11.70	15.02	12.75
X30	60.26	60.67	68.90	94.47	60.60	54.61	52.55	58.71	46.78	49.48	57.19	64.31	55.49	64.47	58.24	64.78
X31	48	49	50	49	49	49	50	49	49	49	49	49	50	49	50	49
X32	5	6	6	6	6	6	6	6	6	6	6	5	6	6	6	6
Under semi-medium land size (2–4 hectares) class																
X1	3.93	3.34	3.77	5.09	4.02	5.24	3.61	4.18	1.46	4.37	3.17	1.49	2.59	2.95	3.98	3.66
X2	2.19	2.49	2.69	2.61	2.76	2.50	2.50	2.79	2.47	2.51	2.68	2.73	2.77	2.79	2.54	2.74
X3	75.76	59.99	61.61	94.71	74.50	67.02	77.34	61.42	99.06	98.85	91.40	78.82	41.74	99.10	55.67	75.38
X4	0.00	0.00	0.00	0.00	0.03	0.22	0.00	0.00	0.00	0.00	0.00	0.10	0.05	0.00	0.00	0.05
X5	96.79	91.52	98.74	62.27	98.26	95.66	99.68	98.74	99.51	99.95	99.20	96.31	98.76	94.91	99.62	93.74
X6	47.31	53.36	50.36	74.43	97.63	66.25	44.91	42.69	70.03	3.67	43.12	12.81	41.80	48.10	90.00	58.35
X7	52.69	46.64	49.64	25.57	2.37	33.75	55.09	57.31	29.97	96.33	56.88	87.19	58.20	51.90	10.00	41.65
X8	48.68	80.25	92.10	28.05	62.96	49.03	91.22	92.98	49.98	7.58	92.60	53.50	34.69	73.61	63.13	59.93
X9	55.42	0.43	59.27	90.09	59.50	39.87	0.07	0.01	15.30	2.53	1.35	11.79	13.66	14.01	5.99	29.32
X10	0	1	5	0	0	1	6	0	0	0	0	0	2	0	1	1
X11	6	0	6	4	0	7	4	1	8	0	6	1	4	0	5	7
X12	36.95	56.98	79.17	52.30	50.56	41.59	90.74	78.53	49.44	7.57	76.29	62.88	28.66	82.54	73.97	59.29
X13	63.05	43.02	20.83	47.70	49.44	58.41	9.26	21.47	50.56	92.43	23.71	37.12	71.34	17.46	26.03	40.71
X14	76.26	75.40	76.85	79.14	80.85	82.91	74.90	76.91	72.45	76.16	72.66	72.39	72.10	75.88	75.49	76.02
X15	68.09	61.47	72.58	90.22	83.32	89.12	57.60	73.06	78.85	67.27	49.44	38.36	64.94	65.19	62.14	68.11
X16	3.07	1.24	1.57	2.09	2.48	2.95	1.12	1.58	0.56	1.41	0.61	0.55	0.48	1.35	1.26	1.49
X17	0.45	0.38	0.49	0.65	0.77	0.92	0.35	0.49	0.17	0.44	0.19	0.17	0.15	0.42	0.39	0.43

(continued)

Vr.	Bam.	Ch-I	Ch-II	E.B.	Gaz.	Hbb.	HCP-I	HCP-II	KL-I	KL-II	KL-III	Manik.	Odl M	Rat.-I	Rat.II	District
X18	0.53	0.46	0.58	0.77	0.92	1.09	0.41	0.58	0.21	0.52	0.23	0.20	0.18	0.50	0.46	0.51
X19	1.73	1.49	1.89	2.53	3.00	3.57	1.35	1.91	0.68	1.70	0.74	0.66	0.58	1.63	1.52	1.66
X20	0.42	0.36	0.46	0.61	0.72	0.86	0.33	0.46	0.16	0.41	0.18	0.16	0.14	0.39	0.37	0.40
X21	10	10	14	9	19	21	10	14	2	6	6	4	3	10	9	10
X22	4	4	6	4	8	9	4	6	1	3	2	2	1	4	4	4
X23	6	6	8	5	10	12	5	8	1	4	3	2	2	6	5	5
X24	56.45	56.42	76.42	50.95	91.61	95.12	52.09	75.11	83.36	65.22	31.11	20.06	75.68	55.84	48.00	62.23
X25	64.35	64.33	76.50	81.00	91.83	100.04	61.70	75.70	88.13	81.43	78.93	82.20	79.54	63.97	59.20	76.59
X26	242	241	327	218	435	493	223	321	257	151	233	386	267	339	205	289
X27	178	178	241	160	320	262	199	236	198	255	291	231	249	276	151	228
X28	90.36	93.36	86.59	98.51	90.32	80.92	90.13	93.77	93.35	90.51	93.57	91.03	93.70	93.45	94.54	91.61
X29	9.64	6.64	13.41	1.49	9.68	19.08	9.87	6.23	6.65	9.49	6.43	8.97	6.30	6.55	5.46	8.39
X30	75.15	73.43	73.11	64.35	57.54	55.53	66.58	63.97	53.38	63.70	56.20	52.52	48.58	60.02	74.63	61.91
X31	51	52	52	52	52	52	52	52	52	52	52	53	53	52	52	52
X32	6	6	6	7	7	7	7	6	6	6	6	6	6	6	6	6
Under medium land size (4–10 hectares) class																
X1	0.57	0.28	0.15	0.52	0.28	1.16	0.05	0.05	0.03	0.03	0.15	0.06	0.15	0.15	0.10	0.29
X2	5.78	5.83	4.74	5.85	4.32	5.56	4.45	4.25	4.50	4.00	5.10	5.31	4.58	5.24	4.48	5.38
X3	93.93	43.87	98.28	96.56	97.97	83.58	95.00	97.54	95.00	96.75	74.42	97.35	65.11	99.03	98.28	86.79
X4	0.00	0.00	0.00	0.00	0.00	0.18	0.00	0.00	0.00	0.00	0.32	0.00	0.00	0.00	0.00	0.05
X5	82.75	94.33	99.45	62.15	98.58	97.60	96.02	99.73	92.07	60.41	99.77	92.83	84.67	89.02	97.58	89.75
X6	73.72	32.41	73.85	72.78	90.76	55.35	61.22	3.92	96.73	95.14	29.15	11.45	65.05	22.18	66.32	60.40
X7	26.28	67.59	26.15	27.22	9.24	44.65	38.78	96.08	3.27	4.86	70.85	88.55	34.95	77.82	33.68	39.60
X8	60.83	25.84	97.87	11.77	58.25	55.99	92.75	98.32	46.80	58.66	97.75	6.05	76.79	82.82	95.64	52.80
X9	44.52	0.50	61.94	67.65	41.94	44.65	0.16	0.11	0.81	0.06	0.34	0.75	76.55	19.21	0.99	37.96
X10	4	3	12	2	0	0	2	0	4	62	0	2	8	1	16	3

Vr.	Bam.	Ch-I	Ch-II	E.B.	Gaz.	Hbb.	HCP-I	HCP-II	KL-I	KL-II	KL-III	Manik.	Odl M	Rat.-I	Rat.II	District
X11	3	1	5	5	0	6	1	1	1	0	4	2	0	1	6	7
X12	56.20	30.84	65.45	17.45	51.97	52.83	76.38	85.58	51.71	97.46	64.40	10.19	64.36	90.59	91.90	54.14
X13	43.80	69.16	34.55	82.55	48.03	47.17	23.62	14.42	48.29	2.54	35.60	89.81	35.64	9.41	8.10	45.86
X14	80.55	75.17	73.13	80.69	78.66	82.95	70.80	70.87	70.58	70.44	71.45	71.16	71.38	73.35	71.67	74.19
X15	68.35	74.33	72.89	90.81	86.37	83.19	63.53	74.76	84.84	77.38	54.60	59.68	91.37	76.58	68.29	75.13
X16	7.25	1.43	0.87	2.97	2.40	9.15	0.22	0.24	0.16	0.12	0.40	0.32	0.38	0.93	0.46	1.82
X17	1.38	0.67	0.41	1.40	1.13	4.30	0.10	0.11	0.08	0.06	0.19	0.15	0.18	0.44	0.22	0.72
X18	1.72	0.84	0.51	1.75	1.42	5.39	0.13	0.14	0.10	0.07	0.24	0.19	0.23	0.55	0.27	0.90
X19	4.15	2.03	1.23	4.20	3.40	12.95	0.31	0.34	0.23	0.17	0.57	0.46	0.54	1.31	0.66	2.17
X20	0.00	0.00	0.00	0.00	0.00	0.00	0.00	0.00	0.00	0.00	0.00	0.00	0.00	0.00	0.00	0.00
X21	12	4	4	5	7	32	1	1	0	0	2	1	1	4	2	5
X22	5	2	2	2	3	14	0	0	0	0	1	0	1	2	1	2
X23	7	3	3	3	5	20	1	1	0	0	1	0	1	2	1	3
X24	64.53	39.07	35.18	47.19	66.64	84.32	58.88	68.54	84.23	71.61	47.25	56.93	73.45	33.24	16.83	56.53
X25	76.36	66.02	62.43	73.50	91.43	92.10	38.19	37.87	93.90	81.48	75.90	86.39	72.40	60.64	45.51	70.27
X26	407	152	137	184	459	406	235	333	299	336	267	227	252	429	265	293
X27	327	321	321	209	309	261	201	227	200	215	259	261	238	304	253	260
X28	97.36	97.36	97.40	99.39	97.53	97.46	92.50	97.44	98.36	97.39	97.40	99.48	97.42	97.38	99.39	97.55
X29	2.64	2.64	2.60	0.61	2.47	2.54	7.50	2.56	1.64	2.61	2.60	0.52	2.58	2.62	0.61	2.45
X30	53.36	75.10	45.48	75.48	65.86	59.13	61.63	62.69	58.46	66.35	51.15	56.92	60.10	68.65	54.33	62.31
X31	52	52	52	52	52	52	52	52	52	52	52	52	52	52	52	52
X32	6	6	5	5	6	6	6	6	6	6	6	6	5	6	6	6

(continued)

Vr.	Bam.	Ch-I	Ch-II	E.B.	Gaz.	Hbb.	HCP-I	HCP-II	KL-I	KL-II	KL-III	Manik.	Odl M	Rat-I	Rat.II	District
Under large land size (>10 hectares) class																
X1	0.02	N.A	N.A	N.A	N.A	N.A	N.A	N.A	N.A	N.A	0.05	N.A	N.A	N.A	N.A	0.00
X2	10.77	N.A	N.A	N.A	N.A	N.A	N.A	N.A	N.A	N.A	10.00	N.A	N.A	N.A	N.A	11.00
X3	95.00	N.A	N.A	N.A	N.A	N.A	N.A	N.A	N.A	N.A	95.00	N.A	N.A	N.A	N.A	97.77
X4	0.00	N.A	N.A	N.A	N.A	N.A	N.A	N.A	N.A	N.A	0.00	N.A	N.A	N.A	N.A	0.00
X5	62.49	N.A	N.A	N.A	N.A	N.A	N.A	N.A	N.A	N.A	87.66	N.A	N.A	N.A	N.A	75.84
X6	95.00	N.A	N.A	N.A	N.A	N.A	N.A	N.A	N.A	N.A	95.00	N.A	N.A	N.A	N.A	59.69
X7	5.00	N.A	N.A	N.A	N.A	N.A	N.A	N.A	N.A	N.A	5.00	N.A	N.A	N.A	N.A	5.87
X8	58.36	N.A	N.A	N.A	N.A	N.A	N.A	N.A	N.A	N.A	57.48	N.A	N.A	N.A	N.A	43.43
X9	0.00	N.A	N.A	N.A	N.A	N.A	N.A	N.A	N.A	N.A	0.00	N.A	N.A	N.A	N.A	0.00
X10	21	N.A	N.A	N.A	N.A	N.A	N.A	N.A	N.A	N.A	0	N.A	N.A	N.A	N.A	9
X11	0	N.A	N.A	N.A	N.A	N.A	N.A	N.A	N.A	N.A	4	N.A	N.A	N.A	N.A	2
X12	95.75	N.A	N.A	N.A	N.A	N.A	N.A	N.A	N.A	N.A	83.09	N.A	N.A	N.A	N.A	59.16
X13	4.25	N.A	N.A	N.A	N.A	N.A	N.A	N.A	N.A	N.A	16.91	N.A	N.A	N.A	N.A	40.84
X14	80.00	N.A	N.A	N.A	N.A	N.A	N.A	N.A	N.A	N.A	78.33	N.A	N.A	N.A	N.A	79.17
X15	49.35	N.A	N.A	N.A	N.A	N.A	N.A	N.A	N.A	N.A	71.32	N.A	N.A	N.A	N.A	60.34
X16	0.00	N.A	N.A	N.A	N.A	N.A	N.A	N.A	N.A	N.A	0.00	N.A	N.A	N.A	N.A	0.00
X17	0.00	N.A	N.A	N.A	N.A	N.A	N.A	N.A	N.A	N.A	0.00	N.A	N.A	N.A	N.A	0.00
X18	0.00	N.A	N.A	N.A	N.A	N.A	N.A	N.A	N.A	N.A	0.00	N.A	N.A	N.A	N.A	0.00
X19	0.00	N.A	N.A	N.A	N.A	N.A	N.A	N.A	N.A	N.A	0.00	N.A	N.A	N.A	N.A	0.00
X20	0.00	N.A	N.A	N.A	N.A	N.A	N.A	N.A	N.A	N.A	0.00	N.A	N.A	N.A	N.A	0.00
X21	12	N.A	N.A	N.A	N.A	N.A	N.A	N.A	N.A	N.A	3	N.A	N.A	N.A	N.A	8
X22	12	N.A	N.A	N.A	N.A	N.A	N.A	N.A	N.A	N.A	3	N.A	N.A	N.A	N.A	8
X23	20	N.A	N.A	N.A	N.A	N.A	N.A	N.A	N.A	N.A	6	N.A	N.A	N.A	N.A	13
X24	68.57	N.A	N.A	N.A	N.A	N.A	N.A	N.A	N.A	N.A	59.05	N.A	N.A	N.A	N.A	63.81
X25	92.42	N.A	N.A	N.A	N.A	N.A	N.A	N.A	N.A	N.A	82.89	N.A	N.A	N.A	N.A	87.66

Vr.	Bam.	Ch-I	Ch-II	E.B.	Gaz.	Hbb.	HCP-I	HCP-II	KL-I	KL-II	KL-III	Manik.	Odl M	Rat.-I	Rat.II	District
X26	462	N.A	N.A	N.A	N.A	N.A	N.A	N.A	N.A	N.A	362	N.A	N.A	N.A	N.A	412
X27	386	N.A	N.A	N.A	N.A	N.A	N.A	N.A	N.A	N.A	341	N.A	N.A	N.A	N.A	363
X28	99.51	N.A	N.A	N.A	N.A	N.A	N.A	N.A	N.A	N.A	99.75	N.A	N.A	N.A	N.A	52.98
X29	0.49	N.A	N.A	N.A	N.A	N.A	N.A	N.A	N.A	N.A	0.25	N.A	N.A	N.A	N.A	0.37
X30	75.00	N.A	N.A	N.A	N.A	N.A	N.A	N.A	N.A	N.A	65.00	N.A	N.A	N.A	N.A	67.00
X31	54	N.A	N.A	N.A	N.A	N.A	N.A	N.A	N.A	N.A	54	N.A	N.A	N.A	N.A	54
X32	7	N.A	N.A	N.A	N.A	N.A	N.A	N.A	N.A	N.A	7	N.A	N.A	N.A	N.A	7
Under all land size classes																
X1	100.00	100.00	100.00	100.00	100.00	100.00	100.00	100.00	100.00	100.00	100.00	100.00	100.00	100.00	100.00	100.00
X2	0.80	0.73	0.77	0.84	0.78	0.91	0.73	0.75	0.55	0.68	0.69	0.72	0.72	0.74	0.75	0.75
X3	70.43	69.11	35.47	83.99	75.54	51.47	61.57	42.03	97.96	85.86	68.67	81.18	48.82	87.66	45.09	66.75
X4	1.83	0.22	0.02	0.16	0.09	4.73	4.55	1.35	0.01	5.47	3.63	0.05	5.77	0.30	0.00	1.67
X5	96.23	96.97	97.07	74.48	98.47	95.45	99.68	98.18	97.88	99.61	97.98	93.26	97.28	94.59	99.46	95.26
X6	67.66	72.12	71.66	84.38	97.89	74.07	58.02	74.87	92.57	53.79	56.59	59.83	70.82	63.90	94.95	74.78
X7	32.34	27.88	28.34	15.62	2.11	25.93	41.98	25.13	7.43	46.21	43.41	40.17	29.18	36.10	5.05	25.22
X8	54.28	86.16	92.73	30.70	53.13	42.80	96.61	94.78	40.05	38.41	88.64	58.54	43.74	76.19	76.77	63.09
X9	59.43	1.38	55.94	86.31	60.42	45.35	0.90	0.54	15.12	8.43	1.85	5.33	28.46	6.54	6.72	25.48
X10	1	1	6	18	0	1	4	0	1	1	1	0	18	3	4	3
X11	6	1	10	7	2	10	15	2	19	19	6	9	14	7	18	11
X12	52.06	61.45	81.31	50.16	42.22	39.37	92.39	95.69	44.86	39.46	71.35	60.59	42.93	83.04	79.25	61.77
X13	47.94	38.55	18.69	49.84	57.78	60.63	7.61	4.31	55.14	60.54	28.65	39.41	57.07	16.96	20.75	38.23
X14	53.94	61.01	75.62	75.56	58.36	57.63	53.20	65.12	76.21	52.36	60.60	69.99	71.52	76.18	74.27	65.44
X15	53.82	54.30	58.51	58.10	77.25	72.27	48.75	55.06	55.68	49.88	37.79	54.14	61.24	62.32	49.26	56.56
X16	1.51	1.35	1.52	1.50	2.25	2.06	1.13	1.38	1.40	1.18	0.70	1.34	0.68	1.67	1.15	1.39
X17	0.18	0.18	0.20	0.20	0.30	0.27	0.15	0.18	0.19	0.16	0.09	0.18	0.09	0.22	0.15	0.18
X18	0.07	0.07	0.08	0.08	0.12	0.11	0.06	0.07	0.07	0.06	0.04	0.07	0.04	0.09	0.06	0.07

(continued)

Vr.	Bam.	Ch-I	Ch-II	E.B.	Gaz.	Hbb.	HCP-I	HCP-II	KL-I	KL-II	KL-III	Manik.	Odl M	Rat.-I	Rat.II	District
X19	1.00	1.01	1.14	1.12	1.69	1.54	0.85	1.03	1.05	0.88	0.53	1.01	0.51	1.25	0.86	1.03
X20	0.28	0.29	0.32	0.32	0.48	0.44	0.24	0.29	0.30	0.25	0.15	0.28	0.14	0.35	0.24	0.29
X21	31	36	43	26	51	49	35	37	16	17	18	28	11	38	29	31
X22	14	16	19	12	23	22	16	17	7	8	8	13	5	17	13	14
X23	14	16	19	12	23	22	16	17	7	8	8	13	5	17	13	14
X24	60.52	69.70	83.81	50.21	100.26	94.86	68.04	73.01	72.12	62.70	35.28	55.65	70.83	74.96	56.09	68.54
X25	60.27	64.86	71.92	72.11	80.14	77.45	64.03	66.51	86.06	79.36	70.65	81.84	70.42	67.49	58.05	71.41
X26	292	337	405	242	484	458	329	353	155	158	270	269	211	362	271	306
X27	219	253	304	182	291	249	276	286	249	246	256	264	218	239	247	252
X28	31.34	20.55	23.40	7.73	35.20	23.09	20.94	26.05	4.48	7.99	11.92	9.27	19.46	22.24	17.64	18.75
X29	68.66	79.45	76.60	92.27	64.80	76.91	79.06	73.95	95.52	92.01	88.08	90.73	80.54	77.76	82.36	81.25
X30	58.45	59.28	66.56	65.84	58.95	60.34	49.69	60.60	61.68	51.65	50.74	59.01	49.79	73.14	50.57	60.42
X31	48	49	49	48	49	49	48	49	49	48	48	49	48	49	49	51
X32	6	5	5	4	5	5	5	5	5	4	5	5	5	5	5	5

Sources: 1. variables 1–13 sources from Agriculture Census, Govt. of India

2. variables 14–27 sources from Input Survey, Govt. of India

3. variables 28–32 sources from District Statistical Handbook, Malda & Census of India website

Annexure 3h: Output variables of agricultural development in 1995–1996 and 2015–2016

Year	Land size Gr.	Bam.	Ch-I	Ch-II	E.B.	Gaz.	Hbb.	HCP-I	HCP-II	KL-I	KL-II	KL-III	Manik.	Odl M	Rat.-1	Rat.II	District
		Gibbs-Martin Crop Diversification Index															
1995–1996	Marginal	0.476	0.542	0.516	0.657	0.52	0.281	0.463	0.511	0.765	0.754	0.817	0.843	0.41	0.754	0.381	0.632
	Small	0.467	0.543	0.5	0.622	0.466	0.285	0.413	0.492	0.694	0.728	0.761	0.753	0.346	0.647	0.384	0.527
	Semi-medium	0.383	0.529	0.536	0.67	0.448	0.357	0.31	0.491	0.817	0.736	0.796	0.79	0.289	0.766	0.276	0.566
	Medium	0.449	0.453	0.469	0.598	0.412	0.339	0.409	0.644	0.613	0.796	0.788	0.726	0.302	0.776	0.464	0.606
	Large	0.107	N.A	N.A	0.147	0.212	0.087	0	0	0.071	N.A	0.652	N.A	0.037	0	N.A	0.191
	All classes	0.377	0.474	0.426	0.531	0.392	0.272	0.31	0.364	0.594	0.592	0.6	0.565	0.27	0.576	0.23	0.442
2015–2016	Marginal	0.534	0.685	0.76	0.77	0.605	0.449	0.525	0.508	0.82	0.772	0.862	0.872	0.235	0.797	0.687	0.726
	Small	0.497	0.636	0.735	0.748	0.554	0.435	0.492	0.529	0.84	0.539	0.655	0.816	0.131	0.769	0.691	0.694
	Semi-medium	0.462	0.547	0.647	0.739	0.57	0.431	0.408	0.397	0.67	0.679	0.668	0.822	0.214	0.736	0.72	0.64
	Medium	0.462	0.576	0.633	0.553	0.608	0.35	0.513	0.686	0.763	0.388	0.665	0.204	0.703	0.721	0.493	0.561
	Large	0.432	N.A	N.A	N.A	N.A	N.A	N.A	N.A	N.A	N.A	0.773	N.A	N.A	N.A	N.A	0.766
	All classes	0.512	0.632	0.74	0.761	0.587	0.431	0.498	0.5	0.829	0.757	0.687	0.812	0.214	0.785	0.694	0.706

| Year | Land size Gr. | Productivity (Yang's Yield Index) | | | | | | | | | | | | | | | |
		Bam.	Ch-I	Ch-II	E.B.	Gaz.	Hbb.	HCP-I	HCP-II	KL-I	KL-II	KL-III	Manik.	Odl M	Rat.-I	Rat.II	District
1995–1996	Marginal	152.9	152.13	150.32	155.63	146.36	153.6	153.96	158.65	138.96	141.25	150.57	157.86	156.33	153.25	156.35	153.24
	Small	153.65	150.21	148.65	155.7	146.36	154.56	150.24	156.37	140.32	140.26	152.35	156.36	155.35	152.35	155.32	152.57
	Semi-medium	149.65	150.32	149	153.25	143.66	155.25	148.66	157.99	141.25	139.65	148.63	158.96	154.57	151.25	153.25	151.72
	Medium	149.37	149.65	147.65	152.32	144.56	152.32	147.37	155.62	139.33	141.25	147.66	155.63	157.66	152.37	154.33	151.17
	Large	148.36	N.A	N.A	152.37	149.21	156.32	148.66	156.46	138.95	N.A	150.24	N.A	154.26	150.21	N.A	151.87
	All classes	153.28	152.98	149.65	154.56	145.23	154.99	151.35	157.89	139.35	141.37	151.54	156.85	156.35	152.35	155.32	152.9
2015–2016	Marginal	170.27	197.26	199.36	201.7	188.25	206.66	193.26	209.37	191.56	191.26	204.9	215.7	204.9	199.13	207.25	202.11
	Small	174.56	191.56	188.26	188.26	190.26	177.25	193.25	181.33	197.25	181.27	180.26	195.56	201.62	195.95	187.25	195.97
	Semi-medium	182.36	189.36	186.66	187.26	189.36	176.66	195.66	182.65	198.25	182.62	179.21	193.36	201.24	193.36	187.65	196.37
	Medium	171.26	188.26	185.97	186.84	188.26	176.84	191.26	179.21	196.66	180.21	177.25	196.35	198.25	192.62	186.86	194.66
	Large	160.24	N.A	N.A	N.A	N.A	N.A	N.A	N.A	N.A	N.A	190.25	N.A	203.99	N.A	N.A	175.24
	All classes	172.06	189.06	186.68	188.59	190.65	177.25	194.99	182.33	198.25	182.39	180.25	195.66	203.99	195.7	188.9	196.66

Source: computed by author (data from various sources)

Annexures: Chapter 4

Annexure 4a: Crop diversification index under all crops in Malda District, 1995–1996 and 2015–2016

Land classes	Bam.	Ch-I	Ch-II	E.B.	Gaz.	Hbb.	HCP-I	HCP-II	KL-I	KL-II	KL-III	Manik.	Odl M	Rat.-I	Rat.II	District
1995–1996																
Marginal	0.476	0.542	0.516	0.657	0.520	0.281	0.463	0.511	0.765	0.754	0.817	0.843	0.410	0.754	0.381	0.632
Small	0.467	0.543	0.500	0.622	0.466	0.285	0.413	0.492	0.694	0.728	0.761	0.753	0.346	0.647	0.384	0.527
Semi-medium	0.383	0.529	0.536	0.670	0.448	0.357	0.310	0.491	0.817	0.736	0.796	0.790	0.289	0.766	0.276	0.566
Medium	0.449	0.453	0.469	0.598	0.412	0.339	0.409	0.644	0.613	0.796	0.788	0.726	0.302	0.776	0.464	0.606
Large	0.107	N.A	N.A	0.147	0.212	0.087	0.000	0.000	0.071	N.A	0.652	N.A	0.037	0.000	N.A	0.191
All classes	0.377	0.474	0.426	0.531	0.392	0.272	0.310	0.364	0.594	0.592	0.600	0.565	0.270	0.576	0.230	0.442
2015–2016																
Marginal	0.534	0.685	0.760	0.770	0.605	0.449	0.525	0.508	0.820	0.772	0.702	0.802	0.235	0.797	0.687	0.726
Small	0.497	0.636	0.735	0.748	0.554	0.435	0.492	0.529	0.840	0.539	0.655	0.816	0.131	0.769	0.691	0.694
Semi-medium	0.462	0.547	0.647	0.739	0.570	0.431	0.408	0.397	0.670	0.679	0.668	0.822	0.214	0.736	0.720	0.640
Medium	0.462	0.576	0.633	0.553	0.608	0.350	0.513	0.686	0.763	0.388	0.665	0.204	0.703	0.721	0.493	0.561
Large	0.432	N.A	N.A	N.A	N.A	N.A	N.A	N.A	N.A	N.A	0.773	N.A	N.A	N.A	N.A	0.766
All classes	0.512	0.632	0.740	0.761	0.587	0.431	0.498	0.500	0.829	0.757	0.687	0.812	0.214	0.785	0.694	0.706

Annexure 4b: Crop diversification index under cereals in Malda District, 1995–1996 and 2015–2016

1995–1996

Marginal	0.083	0.130	0.165	0.262	0.156	0.042	0.212	0.246	0.425	0.339	0.542	0.641	0.150	0.478	0.191	0.283
Small	0.109	0.141	0.138	0.220	0.140	0.053	0.165	0.235	0.420	0.420	0.508	0.593	0.132	0.467	0.198	0.222
Semi-medium	0.072	0.148	0.155	0.228	0.121	0.053	0.123	0.218	0.466	0.398	0.488	0.534	0.055	0.452	0.117	0.219
Medium	0.217	0.054	0.170	0.163	0.080	0.024	0.068	0.306	0.280	0.518	0.485	0.484	0.064	0.518	0.163	0.271
Large	0.000	N.A	N.A	0.000	0.039	0.000	0.000	0.000	0.000	N.A	0.573	N.A	0.037	0.000	N.A	0.046
All classes	0.106	0.133	0.156	0.231	0.134	0.047	0.178	0.238	0.416	0.405	0.514	0.610	0.120	0.478	0.174	0.250

2015–2016

Marginal	0.121	0.067	0.392	0.127	0.344	0.059	0.157	0.135	0.611	0.299	0.403	0.483	0.020	0.519	0.423	0.331
Small	0.088	0.049	0.364	0.210	0.364	0.047	0.131	0.132	0.493	0.151	0.386	0.584	0.000	0.508	0.412	0.327
Semi-medium	0.057	0.032	0.217	0.182	0.413	0.091	0.103	0.097	0.000	0.545	0.422	0.619	0.062	0.458	0.387	0.271
Medium	0.066	0.064	0.174	0.000	0.419	0.092	0.165	0.522	0.408	0.000	0.310	0.480	0.289	0.495	0.290	0.201
Large	0.227	N.A	N.A	N.A	N.A	N.A	N.A	N.A	N.A	N.A	0.375	N.A	N.A	N.A	N.A	0.406
All classes	0.101	0.058	0.360	0.155	0.362	0.066	0.145	0.129	0.556	0.298	0.402	0.534	0.022	0.511	0.416	0.320

Annexure 4c: Crop diversification index under pulses in Malda District, 1995–1996 and 2015–2016

1995–1996																
Marginal	0.000	0.282	0.596	0.685	0.728	0.516	0.540	0.519	0.529	0.502	0.396	0.587	0.376	0.174	0.468	0.645
Small	0.000	0.070	0.305	0.518	0.610	0.517	0.000	0.656	0.601	0.420	0.233	0.494	0.278	0.500	0.467	0.560
Semi-medium	0.000	0.632	0.343	0.520	0.612	0.035	0.000	0.025	0.604	0.000	0.536	0.436	0.544	0.201	0.498	0.556
Medium	N.C	0.430	N.C	0.475	0.562	0.242	N.C	0.551	0.000	0.310	0.299	0.196	0.000	0.297	0.657	0.578
Large	N.C	N.A	N.A	N.C	N.C	N.C	N.C	N.C	N.C	N.A	N.C	N.A	N.C	N.C	N.A	N.C
All classes	0.000	0.418	0.524	0.657	0.710	0.586	0.606	0.484	0.613	0.405	0.551	0.527	0.371	0.197	0.498	0.609
2015–2016																
Marginal	0.403	0.664	0.000	0.514	N.C	N.C	0.264	0.000	0.290	0.000	0.527	0.236	N.C	0.296	0.000	0.548
Small	0.000	0.643	0.000	0.000	N.C	N.C	0.000	0.000	N.C	N.C	0.341	0.308	N.C	0.000	0.000	0.000
Semi-medium	0.000	0.646	0.000	N.C	N.C	N.C	N.C	0.000	N.C	0.000	0.346	0.000	N.C	0.000	N.C	0.630
Medium	0.278	N.C	N.A	N.C	0.000	N.C	0.500	N.C	0.500	N.C	N.C	0.000	0.438	0.000	N.C	0.568
Large	N.C	N.A	N.A	N.A	N.A	N.A	N.A	N.A	N.A	N.A	N.A	N.A	N.A	N.A	N.A	0.000
All classes	0.226	0.665	0.000	0.495	N.C	0.000	0.223	0.000	0.321	0.000	0.550	0.263	0.375	0.243	0.000	0.558

Annexure 4d: Crop diversification Index under spices in Malda District, 1995–1996 and 2015–2016

1995–1996

Marginal	0.590	0.500	N.C	0.174	0.556	0.095	0.000	0.000	0.000	0.000	0.000	0.000	0.000	0.142	0.000	0.000	0.000
Small	0.505	N.C	N.C	0.000	0.271	0.174	N.C	N.C	N.C	0.375	0.000	0.000	0.000	0.000	N.C	N.C	0.000
Semi-medium	0.428	N.C	0.000	0.000	0.000	0.000	0.000	0.000	0.000	0.000	0.000	0.000	0.000	N.C	0.000	N.C	N.C
Medium	N.C	N.C	N.C	N.C	N.C	N.C	N.C	N.C	N.C	N.C	N.C	N.C	N.C	N.C	N.C	N.C	N.C
Large	N.C	N.A	N.C	N.C	N.A	N.C	N.A	N.C	N.C	N.A	N.A	N.A	N.A	N.C	N.C	N.A	N.C
All classes	0.591	0.500	0.000	0.091	0.554	0.062	0.000	0.000	0.000	0.611	0.000	0.000	0.000	0.024	0.000	0.089	0.000

2015–2016

Marginal	0.588	0.305	0.137	0.000	0.596	0.325	0.000	0.531	0.000	0.622	0.000	0.000	N.C	0.473	0.439	0.668	0.000
Small	0.168	N.C	N.C	N.C	N.C	0.000	N.C	N.C	N.C	0.375	N.C	N.C	N.C	0.142	N.C	N.C	N.C
Semi-medium	0.000	N.C	N.C	N.C	0.000	0.000	N.C	N.C	N.C	N.C	N.C	N.C	N.C	N.C	N.C	N.C	N.C
Medium	N.C	N.C	N.C	N.C	N.C	N.C	N.C	N.C	N.C	N.C	N.C	N.C	N.C	N.C	N.C	N.C	N.A
Large	0.000	N.A	N.A	N.A	N.A	0.000	N.A	N.A	N.A	N.A	N.A	N.A	N.A	N.A	N.A	N.A	N.C
All classes	0.600	0.305	0.137	0.000	0.578	0.320	0.000	0.598	0.000	0.611	0.000	0.000	N.C	0.477	0.439	0.678	0.000

Annexure 4e: Crop diversification index under fruits in Malda District, 1995–1996 and 2015–2016

1995–1996

Marginal	0.000	0.000	0.137	0.009	0.000	0.000	0.000	0.000	0.003	0.232	0.538	0.040	0.375	0.000	0.054
Small	0.000	0.000	0.000	0.147	0.000	0.000	0.000	0.000	0.025	0.000	0.241	0.033	0.000	0.000	0.081
Semi-medium	0.000	0.000	0.000	0.065	0.000	0.000	0.000	0.000	0.213	0.000	0.496	0.000	0.000	0.000	0.082
Medium	0.000	0.219	0.000	0.000	0.000	0.000	0.000	0.000	0.000	0.000	0.297	0.000	0.000	0.000	0.053
Large	N.C	N.A	N.A	N.C	N.C	N.C	0.000	0.000	0.000	N.A	0.000	N.A	0.000	N.A	0.000
All classes	0.000	0.171	0.019	0.068	0.000	0.000	0.000	0.000	0.042	0.100	0.432	0.029	0.375	0.000	0.068

2015–2016

Marginal	0.068	0.230	0.005	0.028	0.000	0.578	0.038	0.793	0.351	0.000	0.563	0.047	0.033	0.000	0.177
Small	0.000	0.000	0.181	0.008	0.000	0.000	0.000	0.000	0.448	0.000	0.000	0.000	N.C	0.000	0.000
Semi-medium	0.000	0.000	0.157	0.000	0.000	N.C	0.460	0.000	0.153	0.000	0.000	0.000	0.000	0.000	0.000
Medium	N.C	0.000	0.000	0.000	N.C	0.000	0.000	0.000	0.000	0.000	0.000	0.000	N.C	0.000	0.000
Large	N.C	N.A	N.A	N.A	N.A	N.A	N.A	N.A	N.A	N.A	N.C	N.A	N.A	N.A	N.C
All classes	0.052	0.010	0.214	0.017	0.000	0.000	0.222	0.079	0.376	0.000	0.186	0.021	0.057	0.000	0.138

Annexure 4f: Crop diversification index under vegetables in Malda District, 1995–1996 and 2015–2016

1995–1996

Marginal	0.342	0.338	0.118	0.226	0.246	0.174	0.355	0.023	0.042	0.028	0.094	0.252	0.223	0.228	0.347	0.237
Small	0.329	0.435	0.195	0.118	0.253	0.190	0.209	0.009	0.000	0.000	0.086	0.267	0.177	0.178	0.000	0.221
Semi-medium	0.391	0.306	0.095	0.142	0.331	0.191	0.257	0.028	0.107	0.000	0.034	0.204	0.296	0.033	0.000	0.235
Medium	0.193	0.278	0.000	0.000	0.320	0.127	0.000	0.126	0.000	0.000	0.000	0.061	0.336	0.040	0.266	0.132
Large	N.C	N.A	N.A	N.C	N.C	N.C	N.C	N.C	N.A	N.C	N.C	N.A	N.C	N.C	N.A	N.C
All classes	0.340	0.362	0.130	0.159	0.275	0.179	0.330	0.024	0.039	0.014	0.066	0.236	0.232	0.142	0.183	0.222

2015–2016

Marginal	0.169	0.320	0.333	0.812	0.216	0.563	0.573	0.649	0.898	0.423	0.770	0.824	0.333	0.516	0.467	0.667
Small	0.000	0.360	0.000	0.732	0.185	0.313	0.333	0.081	0.721	0.000	0.651	0.631	0.283	0.246	0.000	0.345
Semi-medium	0.136	0.000	0.000	0.360	0.032	0.349	0.314	0.000	0.000	0.000	0.442	0.000	0.180	0.000	0.153	0.000
Medium	0.000	N.C	0.133	N.C	0.000	0.473	N.C	N.C	N.C	N.C	0.500	N.C	0.000	N.C	N.C	0.428
Large	N.C	N.A	N.A	N.A	N.A	N.A	N.A	N.A	N.A	N.A	N.C	N.A	N.A	N.A	N.A	N.C
All classes	0.124	0.310	0.252	0.780	0.187	0.497	0.510	0.469	0.942	0.534	0.790	0.854	0.272	0.487	0.351	0.616

Annexure 4g: Crop diversification index under oilseeds in Malda District, 1995–1996 and 2015–2016

1995–1996

Marginal	0.098	0.426	0.012	0.000	0.053	0.392	0.279	0.098	0.335	0.000	0.223	0.117	0.296	0.094	0.000	0.147
Small	0.044	0.499	0.000	0.324	0.213	0.366	0.000	0.000	0.069	0.000	0.426	0.479	0.500	0.269	N.C	0.250
Semi-medium	0.183	0.498	0.000	0.452	0.251	0.108	0.000	0.000	0.498	0.000	0.049	0.491	0.236	0.473	0.000	0.251
Medium	0.012	0.449	0.000	0.000	0.342	0.486	0.000	0.266	0.000	0.000	0.000	0.473	0.415	0.495	0.000	0.287
Large	N.C	N.A	N.A	N.C	N.C	N.C	N.C	N.C	N.C	N.A	N.C	N.A	N.C	N.C	N.A	N.C
All classes	0.091	0.484	0.006	0.000	0.155	0.437	0.317	0.070	0.395	0.000	0.260	0.409	0.428	0.335	0.430	0.222

2015–2016

Marginal	0.012	0.019	0.051	0.016	0.030	0.008	0.070	0.005	0.072	0.000	0.022	0.047	0.000	0.032	0.012	0.021
Small	0.006	0.150	0.049	0.000	0.028	0.008	0.000	0.000	0.000	N.C	0.000	0.062	0.000	0.000	0.000	0.000
Semi-medium	0.003	0.000	0.000	0.000	0.000	0.000	0.000	0.000	N.C	0.000	0.000	0.147	0.000	0.166	0.000	0.014
Medium	0.000	0.000	0.000	0.000	0.000	0.124	0.000	N.C	0.000	0.000	0.000	N.C	0.245	0.000	N.C	0.069
Large	0.000	N.A	N.A	N.A	N.A	N.A	N.A	N.A	N.A	N.A	0.000	N.A	N.A	N.A	N.A	0.000
All classes	0.009	0.012	0.076	0.010	0.025	0.012	0.052	0.005	0.053	0.000	0.017	0.062	0.006	0.038	0.008	0.026

Annexure 4h: Crop diversification index under fiber in Malda District, 1995–1996 and 2015–2016

1995–1996

Marginal	0.000	0.005	0.057	0.000	0.011	0.000	0.033	0.011	0.000	0.000	0.463	0.133	0.261	0.000	0.000	0.104
Small	N.C	0.000	0.038	N.C	0.034	0.000	0.000	0.000	0.000	0.000	0.304	0.142	0.000	0.000	0.000	0.048
Semi-medium	N.C	0.000	0.000	0.000	0.149	0.000	0.162	0.000	0.000	N.C	0.293	0.000	0.000	0.000	0.000	0.000
Medium	N.C	0.000	0.000	N.C	N.C	N.C	0.000	0.000	0.000	0.000	0.474	0.000	N.C	N.C	N.C	0.137
Large	N.C	N.A	N.A	N.C	N.C	N.C	N.C	N.C	N.C	N.A	N.C	N.A	N.C	N.C	N.A	N.C
All classes	0.000	0.006	0.036	0.000	0.044	0.000	0.042	0.004	0.000	0.000	0.405	0.113	0.136	0.000	0.000	0.086

2015–2016

Marginal	0.000	0.014	0.046	0.017	0.026	0.005	0.000	0.000	0.038	0.000	0.010	0.000	0.000	0.004	0.000	0.000
Small	0.000	0.072	0.000	0.000	0.000	0.000	0.000	0.000	0.000	0.000	0.000	0.000	0.000	0.000	0.000	0.000
Semi-medium	N.C	0.000	0.000	0.000	N.C	0.037	0.000	0.000	N.C	0.000	0.000	0.000	0.000	0.000	0.000	0.002
Medium	N.C	0.000	0.000	0.000	0.000	N.C	0.000	0.000	0.000	N.C	0.000	0.000	0.000	0.000	0.000	0.000
Large	N.C	N.A	N.A	N.A	N.A	N.A	N.A	N.A	N.A	N.A	0.000	N.A	N.A	N.A	N.A	0.000
All classes	0.000	0.029	0.027	0.000	0.020	0.011	0.000	0.000	0.023	0.000	0.006	0.000	0.000	0.003	0.000	0.007

Annexure 4i: Number of crops under different crops categories in Malda District

1995–1996

Cereals	2	4	4	6	5	4	4	4	5	4	5	5	4	5	5	6
Pulses	2	3	3	3	3	3	3	3	2	2	3	3	3	3	3	3
Spices	1	2	1	2	1	1	1	0	1	1	2	3	2	1	1	3
Fruits	1	2	2	2	1	0	1	1	2	3	3	4	2	2	1	4
Vegetables	8	7	7	9	6	9	8	6	6	9	5	9	8	9	6	9
Oilseeds	1	3	4	5	5	4	3	5	5	4	5	5	4	5	5	5
Fiber	1	4	2	1	2	1	4	2	1	1	3	2	2	1	1	4
Total crops	16	25	23	28	23	22	24	21	22	24	26	31	25	26	22	34

2015–2016

Cereals	5	5	5	4	7	4	5	4	6	3	5	7	3	4	3	7
Pulses	3	5	1	3	2	1	5	2	5	1	5	6	2	5	1	6
Spices	1	6	3	3	1	0	3	3	4	1	4	4	1	2	2	6
Fruits	3	4	6	8	1	1	4	7	8	1	6	9	5	9	1	9
Vegetables	12	15	15	16	9	11	12	9	18	15	15	20	16	17	14	20
Oilseeds	6	3	6	3	4	4	5	2	4	1	4	6	2	3	3	6
Fiber	1	3	2	1	2	2	1	1	2	1	2	1	1	2	1	3
Total crops	31	41	38	38	26	23	35	28	47	23	41	53	30	42	25	57

Source: computed by Author (Data extracted from Agriculture Census, Govt. of India)

Annexure: Chapter 5

Annexure 5a: Macro Nutrients status of Soil in Selected Villages, 2016–17

Villages / Soil quality		Khutadaha	Hatinda	Hazaratpur	Jot Basanta	Bade Mayna	Dalla	Rangaipur	Uttar Kumedpur	Sultanganj	Panchanandapur	Par Deonapur	Mabarakpur	Chandipur	Gadai Maharajpur	Pukhuria	District
Soil pH	Highly acidic	0.00	0.00	0.00	0.00	0.00	0.00	0.00	0.00	0.00	0.00	0.00	0.00	0.00	0.00	0.00	0.00
	Strongly acidic	0.00	0.00	0.00	0.00	8.80	0.00	0.00	0.00	0.00	0.00	0.00	0.00	0.00	0.00	0.00	0.54
	Moderately acidic	52.23	0.00	10.45	3.13	80.80	8.63	5.13	0.00	0.00	0.58	0.00	26.00	0.00	1.18	17.32	14.66
	Slightly acidic	16.60	4.76	44.78	15.63	5.60	25.18	20.51	25.45	3.21	6.77	0.00	32.00	7.25	10.59	25.14	13.05
	Neutral	12.96	3.17	8.96	6.25	0.00	0.00	7.69	3.64	1.07	3.87	0.00	8.00	1.45	1.18	9.50	4.59
	Moderately alkaline	18.22	92.06	35.82	75.00	4.80	66.19	66.67	70.91	95.71	88.78	100.00	34.00	91.30	87.06	48.04	67.16
Electrical conductivity	Critical for germination	0.00	0.00	0.00	0.00	0.00	0.00	0.00	0.00	0.00	0.00	0.00	0.00	0.00	0.00	0.00	0.00
	Harmful crop growth	0.00	0.00	0.00	0.00	0.00	0.00	0.00	0.00	0.00	0.00	0.00	0.00	0.00	0.00	0.00	0.00
	Injurious to crops	0.00	0.00	0.00	0.00	0.00	0.00	0.00	0.00	0.00	0.00	0.00	0.00	1.45	0.00	0.00	0.10
	Normal	100.00	100.00	100.00	100.00	100.00	100.00	100.00	100.00	100.00	100.00	100.00	100.00	98.55	100.00	100.00	99.90
Organic carbon	Very low	9.31	20.63	31.34	15.63	9.60	7.91	7.69	9.09	4.29	16.05	6.67	22.00	19.57	7.06	13.48	12.62
	Low	74.49	58.73	61.19	53.13	48.00	71.94	66.67	61.82	40.71	67.50	80.00	58.00	65.22	51.76	51.69	60.68
	Medium	15.38	19.05	7.46	31.25	27.20	20.14	23.08	27.27	48.57	15.28	13.33	16.00	13.04	32.94	30.34	23.37
	High	0.00	1.59	0.00	0.00	13.60	0.00	2.56	1.82	6.07	1.16	0.00	4.00	2.17	8.24	3.93	3.03
	Very high	0.81	0.00	0.00	0.00	1.60	0.00	0.00	0.00	0.36	0.00	0.00	0.00	0.00	0.00	0.56	0.29

Phosporous (P)	Very low	0.00	0.00	0.00	0.00	3.20	0.00	0.00	0.00	0.36	0.00	0.00	0.00	0.00	1.18	0.00	0.29
	Low	1.21	0.00	0.00	0.00	32.00	0.00	0.00	0.00	1.43	3.48	0.00	0.00	0.00	0.00	0.00	3.18
	Medium	19.43	4.76	1.49	18.75	36.80	9.35	7.69	18.18	30.36	55.13	30.00	2.00	10.87	8.24	3.91	26.34
	High	22.27	11.11	13.43	21.88	12.00	23.74	10.26	23.64	25.71	22.24	20.00	6.00	18.84	20.00	0.00	18.67
	Very high	57.09	84.13	85.07	59.38	16.00	66.91	82.05	58.18	42.14	19.15	50.00	92.00	70.29	70.59	96.09	51.52
Potassium (K)	Very low	0.00	0.00	0.00	0.00	1.60	0.00	0.00	0.00	3.57	0.00	0.00	2.00	0.72	0.00	0.00	0.68
	Low	9.72	14.29	7.46	9.38	15.20	12.95	5.13	14.55	8.93	6.58	0.00	12.00	12.32	11.76	8.38	9.53
	Medium	85.43	71.43	83.58	62.50	83.20	58.27	87.18	56.36	63.21	62.28	50.00	82.00	65.22	50.59	88.27	69.79
	High	4.86	14.29	8.96	28.13	0.00	28.78	7.69	29.09	24.29	30.75	46.67	4.00	21.74	37.65	3.35	19.84
	Very high	0.00	0.00	0.00	0.00	0.00	0.00	0.00	0.00	0.00	0.39	3.33	0.00	0.00	0.00	0.00	0.15

Annexure 5b: Micronutrients Status of Soil in Selected Villages, 2016–2017

Villages / Soil quality		Khuta-daha	Hatinda	Hazar-atpur	Jot Basanta	Bade Mayna	Dalla	Rangaipur	Uttar Kumedpur	Sultanganj	Panchana-ndapur	Par Deonapur	Mabar-akpur	Chandipur	Gadai Maharajpur	Pukhuria	District
Copper (Cu)	Sufficient	77.73	84.13	86.57	96.00	81.60	97.12	79.49	80.00	80.51	88.97	100.00	90.00	96.38	88.24	0.00	86.43
	Deficient	22.27	15.87	13.43	4.00	18.40	2.88	20.51	20.00	19.49	11.03	0.00	10.00	3.62	11.76	0.00	13.57
Iron (Fe)	Sufficient	100.00	96.83	100.00	100.00	100.00	100.00	100.00	100.00	90.71	99.81	100.00	100.00	100.00	100.00	49.43	94.22
	Deficient	0.00	3.17	0.00	0.00	0.00	0.00	0.00	0.00	9.29	0.19	0.00	0.00	0.00	0.00	50.57	5.78
Manganese (Mn)	Sufficient	46.56	76.19	77.61	76.00	72.00	69.06	48.72	61.82	65.34	69.25	56.67	56.00	72.46	76.47	0.00	65.81
	Deficient	53.44	23.81	22.39	24.00	28.00	30.94	51.28	38.18	34.66	30.75	43.33	44.00	27.54	23.53	0.00	34.19
Zinc (Zn)	Sufficient	0.40	9.52	7.46	12.50	0.80	0.00	0.00	29.09	11.07	8.70	0.00	0.00	2.90	24.71	6.15	7.09
	Deficient	99.60	90.48	92.54	87.50	99.20	100.00	100.00	70.91	88.93	91.30	100.00	100.00	97.10	75.29	93.85	92.91

Source: Soil Health Card, Govt. of India

Annexure 5c: Input Variables of Agricultural Development Under Marginal Land Size (<1 hectare) Class

Variables	Khutadaha	Hatinda	Hazaratpur	Jot Basanta	Bade Mayna	Dalla	Rangaipur	Uttar Kumedpur	Sultanganj	Panchan-andapur	Par Deonapur	Mabarakpur	Chand-ipur	Gadai Maharajpur	Pukhuria	District
X1	24	24	24	24	24	24	24	24	24	24	24	24	24	24	24	360
X2	0.42	0.34	0.35	0.36	0.43	0.46	0.36	0.34	0.14	0.3	0.32	0.33	0.34	0.35	0.44	0.35
X3	84.23	88.36	77.3	71.3	69	59.4	79.3	81.2	89	77.82	71.3	79.68	71.3	66.6	61.3	75.12
X4	7.38	9.11	1.82	1.26	0	4.11	1.26	0.56	0.85	5.63	2.33	4.56	15.6	3.65	6.54	4.31
X5	84.85	88.56	77.9	100	100	85.6	100	91.2	100	74.56	90.2	78.98	82.6	81.2	100	89.05
X6	40.26	88.92	92.4	70.2	48.9	42.4	100	97.6	66.4	75.25	88.8	72.56	80.1	85.6	87.6	75.79
X7	0	1	1	2	0	2	1	1	0	1	0	2	0	1	1	1
X8	2	1	0	4	2	5	2	3	2	1	2	6	4	2	2	3
X9	50.26	52.89	56	54.1	76.7	67.9	46.3	53	56.9	47.72	34.1	51.72	32.5	61.7	46.3	52.53
X10	75.63	81.23	79	89.6	71.2	71.7	82.6	69.5	91.2	69.23	71.2	77.68	79.6	80.2	88.6	78.55
X11	2	1	4	3	4	4	4	13	5	10	9	3	2	5	6	75
X12	2	1	0	1	2	1	1	1	0	1	2	1	0	1	3	17
X13	5	6	4	8	9	7	6	5	4	6	9	4	8	7	6	94
X14	3	4	5	3	6	3	4	3	4	6	6	3	4	3	5	62
X15	6	7	4	3	6	9	8	7	8	9	6	4	4	5	3	6
X16	75	91.67	87.5	70.8	91.7	66.7	79.2	70.8	91.7	58.33	79.2	58.33	95.8	83.3	66.7	77.78
X17	62.52	74.56	68.3	66.9	79	61.3	85.7	79.7	82.6	77.25	69.8	78.52	82.6	89.9	91.3	76.65
X18	1	0	0	3	1	1	0	1	5	3	2	6	4	2	5	34
X19	65	78.56	69.6	86	71.3	65.3	82.6	79	86.4	65.32	61.3	85.63	61.3	65.5	78.5	73.39
X20	63.55	77.92	66.6	82.1	87.5	67.6	67.8	67.2	83.7	71.75	64.9	51.06	58.9	70.8	72.8	90.26
X21	329	459	401	634	398	379	611	495	669	510	465	601	354	599	610	501
X22	24.35	35.21	39.7	45.6	24.3	45.2	43.2	26.4	29.4	20	52.3	56	16.1	10	12.3	32
X23	75.65	64.79	60.4	54.4	75.7	54.8	56.8	73.7	70.6	80	47.7	44	83.9	90	87.8	68

(continued)

Annexure 5c (continued)

Variables	Khutadaha	Hatinda	Hazaratpur	Jot Basanta	Bade Mayna	Dalla	Rangaipur	Uttar Kumedpur	Sultanganj	Panchan-andapur	Par Deonapur	Maharakpur	Chand-ipur	Gadai Maharajpur	Pukhuria	District
X24	0	0	2	0	0	0	0	0	0	0	0	0	0	3	5	10
X25	21	22	24	26	27	22	24	34	35	46	40	49	47	39	39	33
X26	56.25	61.11	47.1	36.8	27.3	41.2	40	55	24	61.9	37.5	31.58	39.1	35	69.2	44.2
X27	43.75	38.89	52.9	63.2	72.7	58.8	60	45	76	38.1	62.5	68.42	60.9	65	30.8	55.8
X28	62.99	61.81	56.2	63.5	60.4	54.6	52.2	54.9	63.6	69.3	58.5	60.11	60.2	64.7	52.3	59.69
X29	5	0	1	2	0	4	1	0	1	0	1	0	2	0	1	18
X30	1	3	2	5	7	8	1	9	6	1	5	4	0	1	1	33
X31	5	7	5	6	5	6	6	7	4	5	5	6	5	6	5	6
X32	36	37	39	41	42	37	39	49	50	61	55	64	62	54	54	48

Annexure 5d: Input Variables of Agricultural Development Under Small Land Size (1 – 2 hectares) Class

X1	3	3	3	3	3	3	3	3	3	3	3	3	3	3	3	45
X2	1.46	1.54	1.52	1.11	1.12	1.54	1.13	1.12	1.05	1.11	1.56	1.17	1.56	1.16	1.15	1.29
X3	89.23	79.63	71.3	63.3	71.2	50.1	77.9	79.2	87.6	70.12	70.4	70.36	70.3	62.4	60.3	71.55
X4	0	0.56	0.66	0	0	0	1.99	1	0	1.57	1.56	5.62	6.35	5.8	8.94	2.27
X5	85.42	82.12	72.4	91.2	100	84.6	100	79.6	84.2	71.25	82.4	77.23	80.2	80.3	100	84.72
X6	45.26	91.23	94.5	71.3	50.1	43.3	100	99.2	68.6	78.59	90.3	73.25	81.3	86.5	89	77.49
X7	3	2	2	2	1	3	1	2	2	1	1	3	2	0	1	2
X8	4	3	2	4	3	2	3	5	4	2	3	3	4	3	4	3
X9	66.28	53.74	71.8	75.4	98.5	74.9	48.5	58.9	44.9	46.57	32.3	68.88	40.2	69	52.2	60.14
X10	65.47	75.69	70.2	88.9	82.6	76.4	80.3	71.2	95.6	71.25	70.2	80.23	79.6	82.4	87.9	78.53
X11	2	1	2	1	1	0	1	3	1	2	3	1	2	1	1	22
X12	2	2	0	0	3	3	3	3	0	3	3	2	3	0	2	29
X13	3	0	1	2	3	3	3	1	0	1	0	3	0	2	2	24
X14	2	1	2	1	0	2	1	2	1	1	1	2	0	1	2	19
X15	4	5	9	2	7	4	3	4	3	5	4	6	6	7	2	5
X16	33.33	66.66	66.7	66.7	100	33.3	66.7	33.3	33.3	66.6	66.6	33.3	100	66.6	66.6	59.98
X17	88.95	79.56	72.6	79	80.2	69.6	86.4	73.5	83.7	80.21	72.4	80.25	80.6	95.2	95.3	81.16
X18	1	0	1	0	0	1	0	0	0	0	0	0	0	0	0	5
X19	68	77.85	70.2	85.2	70.3	66.3	85.7	79.7	88	64.52	62.4	84.65	63.3	67.9	79.7	74.23
X20	57.93	66.04	74.4	89.2	92.5	75	55.2	62.3	87.8	73.57	66	65.59	62.3	68.3	79.1	81.23
X21	356	487	412	509	432	387	501	512	587	522	467	575	565	512	606	495

(continued)

Annexure 5d (continued)

X22	22.31	50	20	41.2	63	51.3	20	34	43.2	35	32	48	34.3	28	44.2	37.77
X23	77.69	50	80	58.8	37	48.7	80	66	56.8	65	68	52	65.8	72	55.8	62.23
X24	0	0	1	0	0	0	0	0	0	0	0	1	0	1	0	3
X25	42	40	37	43	37	38	29	22	43	22	41	22	35	39	41	35.4
X26	70	66.67	80	100	100	50	77.8	100	100	100	83.3	75	75	100	77.8	83.7
X27	30	33.33	20	0	0	50	22.2	0	0	0	16.7	25	25	0	22.2	16.3
X28	73.15	64.56	62.8	58.4	64.5	58.5	56.5	62.6	70.7	53.37	61.1	68.21	43.4	68.4	62.1	62.19
X29	0	1	0	0	0	1	0	0	2	0	1	0	1	1	1	8
X30	1	2	0	2	0	1	0	0	2	2	1	0	1	1	2	6
X31	5	7	6	5	6	6	7	7	5	6	5	5	6	5	6	6
X32	59	57	54	60	54	55	46	39	60	39	58	39	52	56	58	52

Annexure 5e: Input Variables of Agricultural Development Under Semi-Medium Land Size (2 – 4 hectares) Class

X1	2	2	2	2	2	2	2	2	2	2	2	2	2	2	2	30
X2	2.39	2.29	2.19	2.31	2.26	2.4	2	2.29	2.05	2.01	2.18	2.24	2.27	2.29	2.04	2.21
X3	91.23	77.25	67	61.2	66.3	45.7	76	71.3	82.4	69.32	68.5	66.38	70	60.2	59.9	68.84
X4	2.65	2.98	0	4.62	0	1.55	0	0.56	0	0	0	4.63	0	8.54	2.65	1.88
X5	55.65	80.21	80.2	84.6	91.2	86.3	94.2	77.2	88	70.23	80.2	75.63	86.3	77.6	97.6	81.68
X6	46.32	100	96.4	75.3	52.5	45.6	100	100	69.4	80.25	91.3	65.87	85.6	89	90.2	79.18
X7	4	2	1	2	3	5	4	1	1	2	0	4	1	2	4	2
X8	5	2	1	2	4	2	5	4	3	3	5	2	2	2	5	3
X9	64.08	57.19	68.8	87.1	98.5	87.3	53.2	69.3	33.6	63.23	35.3	33.12	30.8	61.1	57.9	60.02
X10	60.23	70.45	72.4	87.6	77.6	69.4	82.4	69.4	91.3	68.98	69.6	78.96	76.2	79	70.3	74.89
X11	0	1	1	1	1	0	1	2	0	1	2	1	2	1	1	15
X12	1	1	0	0	0	1	1	1	2	1	1	0	0	1	2	12
X13	2	1	2	1	2	2	1	2	1	0	2	1	0	0	1	18
X14	1	1	0	1	1	2	1	2	2	1	2	1	1	2	4	22
X15	3	3	7	2	4	6	3	7	5	2	4	3	4	3	2	4
X16	100	100	100	100	100	50	100	50	100	100	50	100	100	50	100	86.67
X17	100	100	100	100	100	100	100	100	100	100	100	100	100	100	100	100
X18	2	3	2	5	2	3	2	4	3	4	2	2	3	1	1	39
X19	69.32	81.23	71.3	84.6	72.4	64	88	80.2	88.7	66.35	64.6	82.35	64.6	69.5	80.2	75.15
X20	56.45	66.42	76.4	81	81.6	65.1	52.1	75.1	83.4	75.22	63.1	70.06	65.7	55.8	78	69.695
X21	369	491	452	560	452	354	498	505	580	502	487	588	548	568	611	496
X22	45	52	31	38	15	26	34	29	16	46	18	13	47	21	33	30.93
X23	55	48	69	62	85	74	66	71	84	54	82	87	53	79	67	69.07

(continued)

Annexure 5e (continued)

X24	0	0	1	1	0	0	0	0	0	0	0	2	0	1	2	7
X25	34	36	31	30	28	25	21	20	22	31	27	30	33	40	41	29.93
X26	100	100	100	100	100	100	100	100	100	100	100	100	100	100	100	100
X27	0	0	0	0	0	0	0	0	0	0	0	0	0	0	0	0
X28	77.05	68.46	66.7	62.3	68.4	62.4	60.4	66.5	74.6	57.27	65	72.11	47.3	72.3	66	73.33
X29	2	0	0	0	0	1	0	1	2	2	2	0	0	0	1	11
X30	1	1	0	0	0	2	0	0	1	0	1	0	0	2	2	5
X31	6	7	7	6	5	8	8	8	6	7	4	7	4	8	7	7
X32	51	54	49	48	46	43	39	38	40	49	45	48	51	58	59	48

Annexure 5f: Input Variables of Agricultural Development Under Medium Land Size (4 – 10 hectares) Class

X1	1	1	1	1	1	1	1	1	1	1	1	1	1	1	15
X2	4.68	4.73	4.64	4.75	4.22	5.46	4.35	4.15	4.14	4.52	5	5.21	4.48	5.14	4.38	4.66
X3	79.23	72.36	68.3	60.3	67.3	44.2	72.4	70.1	77.3	71.36	65	67.89	67.6	62.5	57.8	66.91
X4	0	0	0	0	0	0	0	0	0	0	0	0	0	0	0	0
X5	78.54	78.56	88.6	79	99.2	80.2	91.2	70.2	80.2	66.35	89.2	74.25	79	76.4	91.2	81.48
X6	45.89	87.69	96.4	77.7	55.5	47.6	100	100	70.1	78.59	92.4	77.45	86.5	100	100	81.05
X7	3	2	1	1	5	4	1	0	1	0	2	6	1	2	2	2
X8	1	3	2	3	2	4	2	2	1	4	1	2	3	1	3	2
X9	96.38	53.31	37	97.5	81.3	75.6	68.4	69	46.7	45.48	23.6	41.3	23.1	38.8	45.4	56.18
X10	61.25	72.32	67.5	86.5	71.3	68.8	81.3	65.5	90.3	68.5	71.2	81.25	74.6	77.3	69.4	73.78
X11	1	1	1	1	1	1	1	1	1	1	1	1	1	1	1	15
X12	0	0	1	1	0	0	1	1	1	0	0	0	1	1	0	7
X13	0	0	1	1	1	0	1	0	1	0	1	1	0	0	1	8
X14	1	2	0	2	2	1	0	0	0	0	0	0	0	1	0	9
X15	5	2	1	3	1	2	1	6	4	3	8	2	7	6	4	4
X16	100	100	100	100	100	100	100	100	100	100	100	100	100	100	100	100
X17	100	100	100	100	100	100	100	100	100	100	100	100	100	100	100	100
X18	2	1	3	1	1	2	0	1	0	2	2	0	0	1	0	16
X19	90	88.56	73.3	85.4	74.5	67.9	89	82.4	89.7	67.54	65.7	88.65	65.8	70.3	81.3	78.65
X20	63.34	62.37	78.2	81.2	72.3	78.3	79.6	69.3	84.6	71.74	68.7	77.52	64.6	66	78.3	73.06
X21	401	499	501	579	411	399	511	489	542	541	471	598	401	499	600	496
X22	45	65	43	43	38	27	16	35	14	32	61	64	16	34	41	38.27

(continued)

Annexure 5f (continued)

	55	35	57	57	62	73	84	65	86	68	39	36	84	66	59	61.73
X23	55	35	57	57	62	73	84	65	86	68	39	36	84	66	59	61.73
X24	0	0	1	0	0	0	0	0	0	0	1	1	0	0	1	4
X25	23	28	46	35	37	50	44	35	21	22	17	20	24	26	28	30.4
X26	100	100	100	100	100	100	100	100	100	100	100	100	100	100	100	100
X27	0	0	0	0	0	0	0	0	0	0	0	0	0	0	0	0
X28	76.22	77.95	78.3	58.3	58.7	62	64.5	65.6	61.3	69.2	64	59.78	63	51.5	57.2	73.33
X29	0	0	0	1	0	0	1	0	1	1	0	0	0	0	1	5
X30	0	0	1	0	0	1	0	1	1	1	0	0	1	1	1	6
X31	7	6	8	7	6	6	8	6	4	5	5	6	5	7	4	6
X32	40	45	63	52	54	67	61	52	38	39	34	37	41	43	45	47

Annexure 5g: Input Variables of Agricultural Development Under All Land Size Classes

X1	30	30	30	30	30	30	30	30	30	30	30	30	30	30	30	450
X2	0.56	0.34	0.48	0.46	0.62	0.62	0.4	0.42	0.22	0.35	0.36	0.39	0.39	0.4	0.42	0.43
X3	85.23	84.22	76.3	68.7	67.8	58.3	80.2	82.4	89.3	78.95	70.7	76.89	68.9	63.9	62.6	70.6
X4	7.38	2.65	2.56	1.26	0	5.64	3.25	2.11	0.85	4.62	1.54	2.98	5.26	2.65	5.63	3.23
X5	85.62	89.23	79.5	91.2	95.6	84.6	96.5	75.6	91.1	75.63	84.6	79.63	83.3	80.3	96.4	85.91
X6	41.23	89.25	94.6	71.3	50.2	43.6	100	98.6	67.9	79.58	90.3	72.51	82.4	94.6	95.6	78.1
X7	3	2	1	2	2	4	2	1	1	1	1	4	1	1	2	2
X8	3	2	1	3	3	3	3	4	3	3	3	3	3	2	4	3
X9	56.47	57.03	62	61.5	83.9	78	50.6	57.9	58.7	51.88	37.8	56.86	37.1	66.4	51.2	57.81
X10	71.23	79.68	76.5	88	74.3	72.5	82.6	68.4	91.6	69.25	70.2	78.52	76.5	79.5	86.6	77.69
X11	5	4	8	6	7	5	7	19	7	14	15	6	7	8	9	127
X12	5	4	1	2	5	5	6	6	3	5	6	3	4	3	7	65
X13	10	7	8	12	15	12	11	8	6	7	12	9	8	9	10	144
X14	7	8	7	7	9	8	6	7	7	8	9	6	5	7	11	112
X15	18	17	21	10	18	21	15	24	20	19	22	15	21	21	10	18
X16	77.08	89.58	88.5	84.4	97.9	62.5	86.5	63.5	81.3	81.23	73.9	72.91	99	75	83.3	81.1
X17	87.87	88.53	85.2	86.5	89.8	82.7	93	88.3	91.6	89.37	85.6	89.69	90.8	96.3	96.6	89.45
X18	5	4	6	9	4	6	2	7	8	9	6	9	7	4	6	94
X19	69.87	79.65	74.5	86.5	72.6	65.9	84.6	79.7	87.6	66.54	62.4	86.32	62.5	66.4	81.2	75.07
X20	64.57	79.85	68.6	83.3	90.4	68.5	69.4	67.8	85	72.49	66.8	56.19	59.8	70.6	73.6	76.945
X21	311	448	411	523	419	374	448	448	561	523	478	602	473	586	619	482
X22	34.17	50.55	33.4	42	35.1	37.4	28.3	31.1	25.6	33.25	40.8	45.25	28.4	23.3	32.6	34.74

(continued)

Annexure 5g (continued)

	65.84	49.45	66.6	58	64.9	62.6	71.7	68.9	74.4	66.75	59.2	54.75	71.7	76.8	67.4	65.26
X23	65.84	49.45	66.6	58	64.9	62.6	71.7	68.9	74.4	66.75	59.2	54.75	71.7	76.8	67.4	65.26
X24	0	0	3	1	0	0	0	0	0	0	2	5	0	2	3	16
X25	28	23	29	22	27	28	30	31	32	29	34	28	26	30	27	28.25
X26	66.67	66.67	66.7	60	46.7	56.7	63.3	70	36.7	73.33	63.3	50	50	56.7	80	60.44
X27	33.33	33.33	33.3	40	53.3	43.3	36.7	30	63.3	26.67	36.7	50	50	43.3	20	39.56
X28	71.9	68.56	62.9	65.2	63.6	60.9	56.9	58.6	65.2	69.9	68.6	70.9	73.1	70.9	71.9	66.6
X29	7	1	1	3	0	6	2	1	6	3	4	0	3	1	4	42
X30	2	1	3	0	0	4	2	1	5	4	2	0	0	5	6	50
X31	6	6	8	7	6	7	8	7	5	6	5	6	5	7	6	6
X32	47	47	48	47	47	47	46	47	47	48	49	47	45	49	46	47

Source: field survey 2018

Index

Printed in the United States
by Baker & Taylor Publisher Services